RECYCLING TWO-LITER CONTAINERS for the TEACHING of SCIENCE

ALFRED DE VITO
PROFESSOR EMERITUS OF SCIENCE EDUCATION
PURDUE UNIVERSITY

CREATIVE VENTURES, INC.
P. O. BOX 2286
West Lafayette, Indiana 47906

Library of Congress Cataloging in Publication Data

De Vito, Alfred

RECYCLING TWO-LITER CONTAINERS FOR THE TEACHING OF SCIENCE

P. O. Box 2286, West Lafayette, Indiana 47906

Current Printing (last digit)

10 9 8 7 6 5 4 3

ISBN: 0-942034-09-0

Printed in the United States of America

FOREWORD

This is a book about the creative use of recycled, two-liter, plastic containers for the teaching of science. Materials for science instruction, at any level, are often expensive and many times not readily available. Also, materials for science instruction are often difficult to store, clean, and maintain. Therefore, it is always exciting to find materials that can alleviate such problems and still promote "good" science. Recycling is not only a productive act, it is a necessary act.

Two-liter, plastic containers are usually plentiful, free, and complement numerous, hands-on science activities. The easy availability of two-liter containers can provide an ample supply of materials to each child engaged in hands-on activities. Every child can become a direct participant rather than a spectator in the arena of science instruction.

AUTHOR'S NOTES, COMMENTS, AND CAUTIONS

Few things in science instruction work well, turn out perfectly, or are so spectacular that they leave students and teachers gasping. Few things work exactly as authors say they do. Never fail, one-hundred percent, accurate, everytime-a-winner activities are almost non-existent. Authors want their suggested activities to work. And, they make them work. You can too. Patience and perseverance are necessary ingredients for success. One attempt to replicate an activity will not necessarily make you an expert. Repeated efforts plus the addition of your own ideas and innovations will move you closer to being an expert. The successful performance of activities hinges on numerous, simple things or conditions not always fully described or even anticipated by authors. A slight temperature difference, a change in humidity, the wrong size cork, an incorrect opening or aperture, etc. can influence the outcome of the activity causing the results to differ from the suggested outcome. These differences are the cornicopia of science instruction. They provide a prosperity of creative thought. They challenge us. Sure-fire never-fail activities deprive the investigator of the creative pleasures associated with searching out one's own solution to problems.

Numerous ideas are presented throughout this book. Many are hands-on activities. Others are suggested extensions into experimental investigations. The replication of activities from this book is the intent of the author. However, it is the engagement and subsequent "fallout" of personal ideas and solutions generated by the investigator, and injected in an embellishing manner, extending, and improving science instruction that evokes the real "VOILAS."

The format of this book is to present an activity, propose some pertinent questions about the activity, and to suggest avenues for further investigations through experimentation. Activities are doorways to the resolution of problems or questions through experimentation. The exploration of all the variables that exist in any given activity far exceeds the space assigned by the author. Authors, like teachers and others, plant seeds of thought. They do not always provide time to fully cultivate the garden. However, if the seeds are good, the garden grows in spite of us.

ABOUT TWO-LITER, PLASTIC CONTAINERS

Not all two-liter, plastic containers are alike. Manufacturers constantly strive to improve their products and this results in variations in the market place. Most two-liter containers are made of clear plastic. Some are color tinted. Bottom sections of containers can vary in structure. Some two-liter containers are manufactured without the added bottom sleeve. Container caps can differ in design and material. Despite all these variables, two-liter containers are extremely useful.

Many activities require punching, cutting, or boring holes in plastic. A few simple tools such as a paper punch, metal or leather punches, awls, scissors, and an assortment of metal punches will enable you to become a "TWO-LITER" expert. Each of the suggested tools has a unique purpose, however, you may find suitable, substitute tools that perform as well or even better. Remember, tools can be dangerous if used improperly. Great care should be taken when using sharp tools that require the application of pressure to cut into or through various materials. Never push sharp tools in a manner that a slip could injure you or someone else. Plastic materials tend to be slippery and care must be exercised when piercing them. When piercing or cutting a two-liter, plastic container, be cognizant of the curvature of the container's surface. Curved surfaces, plus the slipperiness of the plastic, dictate care when applying pressure. Always work slowly, carefully, and in a concentrating manner. No one needs a two-liter injury.

Children should not work with sharp tools.

Do not use heated objects such as a nail to burn a hole through plastic materials. Noxious fumes may be given off. If in doubt about a procedure, obtain help from someone more knowledgeable than you about the materials and the tools involved. Success comes with experience, dedication, and continued application.

CONTENTS

DIFFERENT ORIENTATIONS, DIFFERENT VIEWS.

OTIVeD

SEEDS, PLANTS, AND TWO-LITER POTS

A study of plant growth can provide sufficient avenues of investigation to complement most science programs at any level of instruction. Opportunities for investigating plant growth seem endless. At lower levels of instruction, few materials are needed to get started. These are seeds, water, containers or pots, soil or a substitute material, and light (preferably sunlight).

Two-liter containers can enhance your science instruction because they are readily available; because they are free; and, because they have a multitude of uses. Another added bonus is that they can be discarded for future recycling when one is finished with them, thus requiring little or no cleaning. All this is free!

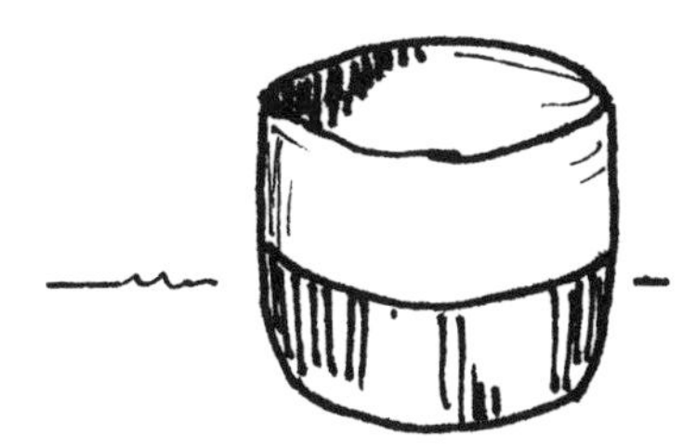

With the top portion of the container removed, the bottom portion of the container makes an excellent, leak proof pot for any use. Pots can be cut out of the container using a utility knife to slice into the plastic container. Once the container has been pierced, using a pair of scissors cut around the container separating the top from the bottom. The height of the pot will be dictated by the requirements of your investigation. Save the top funnel portion and the container cap. These will be useful in other activities.

If you leave the black, bottom section of the container intact, you will have a water-proof planting pot. Of course, this means that any excess watering of seeds or plants will drain and accumulate on the bottom. If this is not desirable, cut a similar pot out of another container. Separate the black section from the container. This can be accomplished by the application of brute force. However, this usually results in a distorted bottom section sometimes rendering it useless. The black section can be removed most easily by the application of heat. This softens the glue and the bottom can easily be separated. A hair dryer/blower works nicely. It is suggested that one direct heat from the hair dryer to the base and lower portion of the two-liter container. This, plus some gentle pulls followed by the application of additional heat, will permit the easy removal of the bottom. Puncture the clear portion of the container. Using an awl, carefully push holes through the container from the inside out to prevent an accident from any slippage. Make as many holes as you deem necessary to accomodate your drainage requirements. Fit this inside the leak-proof pot previously constructed. Good drainage is essential for good plant growth. Most seeds rot from the application of excessive water prior to germination taking place.

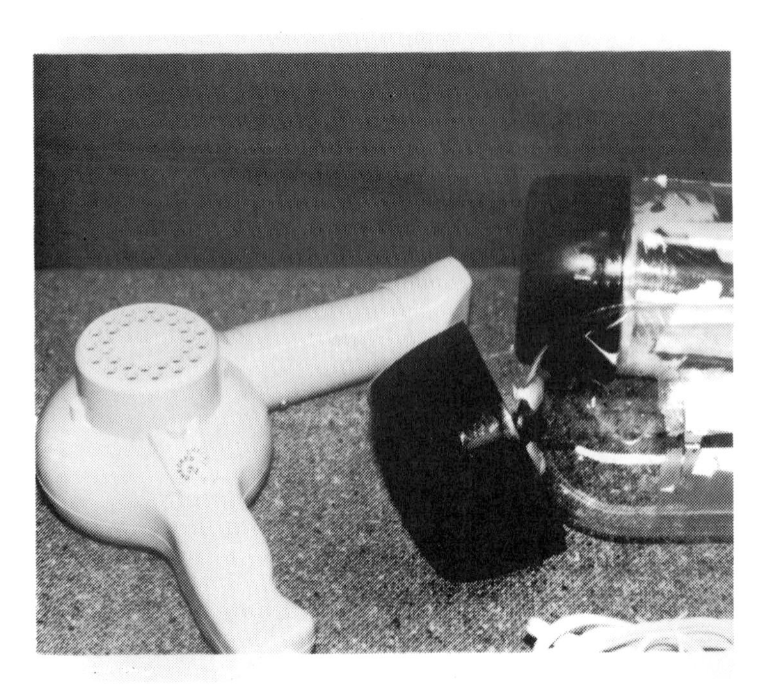

WORKING WITH SEEDS

There are seeds. And, there are seeds. Some are small like mung beans which germinate in about a week. Some are large like lima beans which take about one to two weeks to germinate. Corn seeds are smaller than lima beans and larger than mung beans. These take about three weeks to germinate. Germination is affected by temperature as well as proper watering and sunlight. Young children wish for instantaneous germination. Thus, many teachers elect to grow mung beans as opposed to some other selection. The lima bean is a preferred choice because the plant itself provides larger, more easily examinable parts for measuring, cuttings, and experimentation. By contrast mung beans, while they germinate quickly, grow in a spindling manner. Corn produces a good size plant but its germination time is long.

Seeds can be planted in almost any kind of potting material. Water is prime to plant growth. Soak the seeds for a few hours until they absorb sufficient amounts of water to plump up. This action accelerates the initial growth process. The handling of seeds by many individuals cause seeds to become contaminated. This often times can be prevented by soaking the seeds, for a brief period of time (15-30 minutes), in a household laundry bleach. This can retard any mold growth that might occur. If mold growth does occur on any of your seeds, using tweezers, remove them from your pot. While some plants may emerge from a molding seed, it is best to discard molded seeds rather than contaminate other seeds in the pot. Caution should be taken when working with molds. Use tweezers to remove them and wash your hands carefully afterwards.

PLANT THE SEEDS

When planting your seeds in your two-liter pot, it is best to use seeds of one kind. Mixing and matching seeds for inclusion in one pot confounds things because each seed may have different requirements for growth. One seed may need lots of water and indirect light. By contrast, another seed may need little water and direct light. It becomes difficult to meet all the growth needs of a variety of seeds in one pot. Of course, this suggests an area of investigation.

***** Compare the growth of three, heterogeneous seeds grown in one pot with the growth of seeds in three separate pots each containing homogeneous seeds.

SEED ORIENTATION

Lima beans can be positioned within a planting pot in a variety of ways. Some are: ⟶

A question begging an experiment.

***** Does seed orientation effect growth?

PLANTED DOWN

PLANTED UP

PLANTED VERTICALLY

PLANTED HORIZONTALLY

DEPTH OF SEED PLANTING

Seeds are effected by the depth of planting. Depth of planting requirements vary from seed to seed. Any seed planted too deeply will not grow.

Some seeds will grow if they are simply scattered on the surface of certain materials. If the proper light, water, and temperature are available, seeds grow. A wet sponge, set in a pot and sprinkled with grass and/or bird seed will support growth.

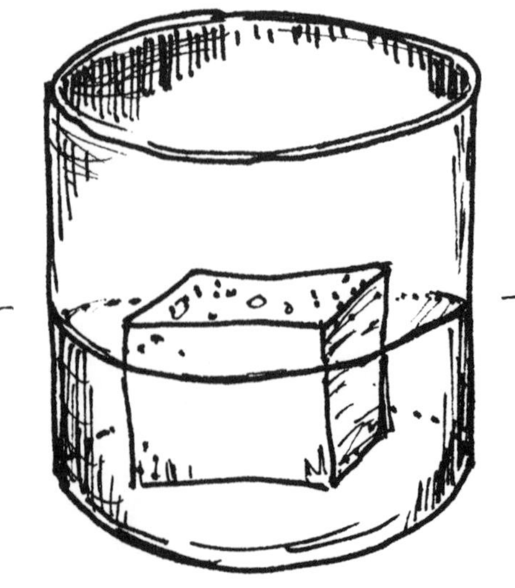

I WANNANO:

***** What is the best depth of planting for lima bean, plant growth?

POTTING MATERIALS

Seeds will grow in almost any material. In which material do seeds grow best? If you elect to investigate the benefits of soil on seed growth, you could plant seeds in loam, sand, or clay and compare the results. The manipulated variable would be the soil and the responding variable would be the growth.

***** Do seeds and the resultant plants grow better in loam, sand, or clay? Or, would some combination of these three soils work better?

This investigation could be extended into investigations of the ability of seeds to grow in many different materials, for example, coffee grounds, sawdust, shredded paper, styrofoam, etc. These observations and results can be compared to the prior soil investigation.

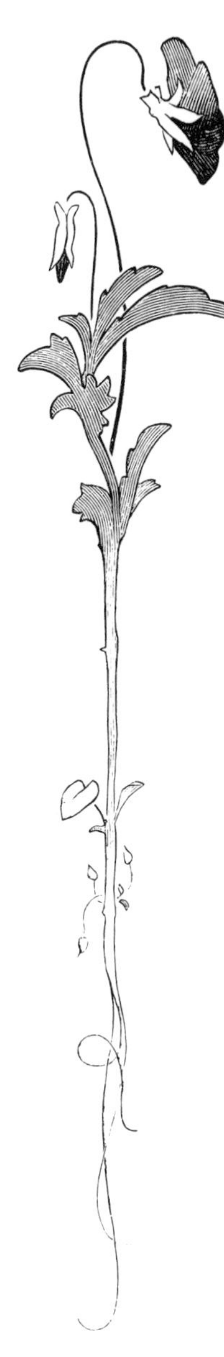

Two-liter containers are tapered at the top and at the bottom. When requirements call for one container to nest inside another container, it is necessary to make cuts into the container's taper. This allows one container to fit snugly into the other container. A cut made at any point on the full cylinder will not permit any other cut cylinder of the same diameter to nest inside of it.

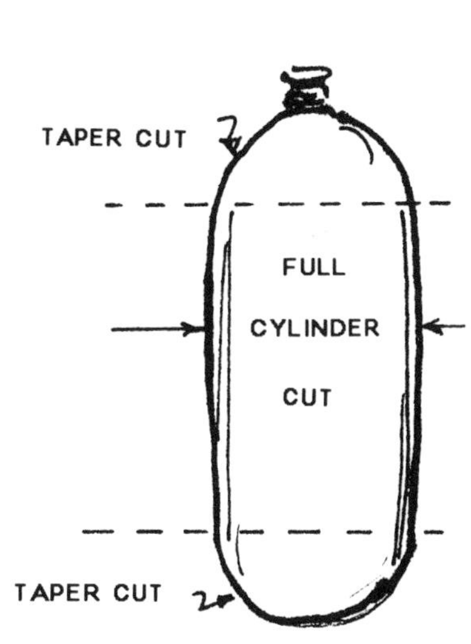

SELF-WATERING POT

A self-watering pot can be constructed from a two-liter container. Cut the lower portion about two-thirds of the container's height. Save the funnel-shaped, top portion. Invert the top portion, minus the cap, and insert it into the bottom portion of the cylinder which should be partially filled with tap water. Using paper toweling, make a wick that will completely wrap around the inside of the inverted plastic funnel and which is long enough to permit it to be pushed down through the neck of the inverted top of the container and extending down into the water reservoir below. Stuff shredded paper into the top portion of the funnel. Initially, saturate the shredded paper with water. Then squeeze out the excess water. This can be done with the funnel removed. Orient your seeds in the funnel and insert this back into the cylinder making sure the paper-towel wick hangs pendant in the water.

If everything goes as designed, the paper wick, through capillary action, will bring water up into the pre-moistened, shredded paper. In the early stages pay particular attention to the wetness of the shredded paper. Excess water will promote rot, mold, or both. The funnel can be separated from the bottom portion at any time to permit a "drying" out period.

Materials other than shredded paper can be used. If soil is used, some device such as a paper-towel plug will be needed to retain the soil in the upper portion of the funnel.

When inserting this plug, be sure that you do not block the capillary action of the wick.

Some suggested variables that may be manipulated are:

- the wick

  Which material makes the best wick?

- the planting material

  Which potting material best promotes plant growth, for example, soil, shredded paper, cotton, etc?

- the liquid

  Which liquid promotes plant growth, for example, tap water, rain water, saltwater, distilled water, etc.?

HYPOTHESIZING AND EXPERIMENTATION

Experimentation in science is the testing of hypotheses. Hypotheses are generalized statements that rise from observations about relationships between variables. Sometimes a hypothesis has been described as an "Educated" guess about prior observations. Usually the hypothesis is stated in an "If... then" statement or a statement to be "disproved."

In experimentation, those variables that will influence the outcome of the experiment must be controlled. This is necessary so that the assignment and subsequent credit for an action can be attributed exclusively to that one variable which is manipulated and to no other existing variable. More than one manipulated variable confounds the experimenter, for no accurate match of cause (manipulated variable) and effect (responding variable) can be assigned to one specific variable. More specifically, variables can be identified as follows:

CONTROLLED VARIABLES - Variables that are held constant through the experiment so as not to influence the results.

MANIPULATED VARIABLE (CAUSE) - - The manipulated variable (sometimes called the independent variable) is that variable purposefully changed in an experiment.

RESPONDING VARIABLE (EFFECT) - - - The responding variable (sometimes called the dependent variable) is that variable that changes in response to the manipulated variable.

AN EXAMPLE:

LIMA BEANS PLANTED IN SOIL WILL GROW FASTER THAN LIMA BEANS PLANTED IN SAWDUST MATERIAL.

The potting material is the manipulated variable. The responding variable is the anticipated differences in growth in response to this manipulation. All other variables are controlled.

EXPERIMENTATION

INVOLVES THE FOLLOWING:

- the statement of a problem usually in the form of a hypothesis
- identification of variables that have a bearing on the stated hypothesis
- the formulation of strategies compatible with the stated hypothesis
- controlling the variables
- collecting and interpreting the data
- the conclusions, and
- the replication of the experiment to validate prior conclusions

SAMPLE INVESTIGATIONS

***** Using two, two-liter pots, compare the growth of a growing stationary plant to one that is positioned on a rotating turntable (Geotrophism)).

***** Will polluted water (detergent water) effect seed germination?

***** Will a specific color of light reflected onto a growing plant effect growth?

PICK AN AREA OF INVESTIGATION

ACTION:

- Formulate a hypothesis for the investigation
- Identify the controlled, manipulated, and responding variables
- Conduct the experiment

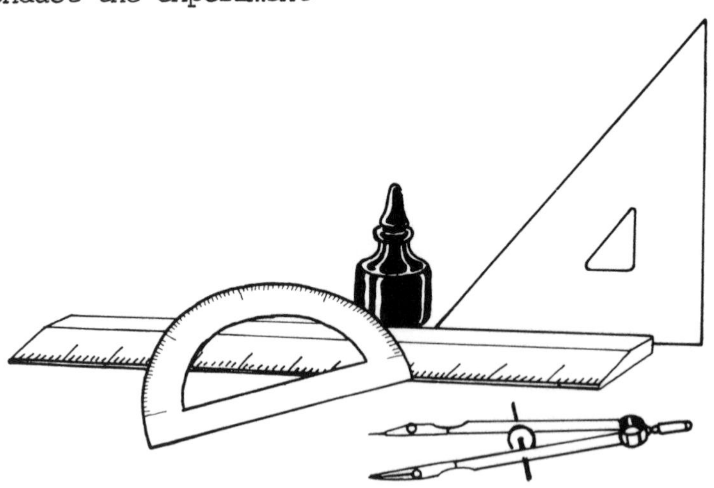

TWO-LITER POTS, PLANTS, AND LIGHT

Light is vital for plant growth. Seeds and plants will germinate and grow in darkness. However, they will not flourish with the same vigor as those exposed to light, particularly natural light.

Not all two-liter containers are made of clear plastic. Some are fabricated with a green tint. For this particular activity this is a plus.

Construct two, two-liter pots.
One clear pot and one green-tinted pot.
Save the tops for later usage.

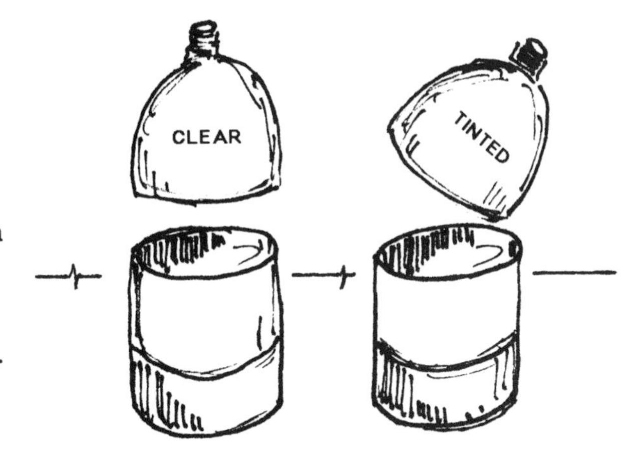

Make an inference about the growth of lima bean seeds planted in each of these containers. State a testable hypothesis regarding growth as effected by light. Translate your hypothesis into action through the performance of an experiment.

- What is your manipulated variable?
- What is the anticipated response variable?

EXTENSIONS...

***** Using two, clear, two-liter pots, compare the growth rate of one, seeded pot in darkness with one seeded pot in sunlight.

- Formulate a hypothesis
- What are the controlled, manipulated, and response variables?
- Conduct an action experiment

***** Using several two-liter pots (two clear, two green, or one of each). Compare plant growth when pots are placed under fluorescent light. Incandescent light. And/or direct light versus indirect light.

- Formulate a hypothesis
- Keep your manipulated variables uncluttered

TWO-LITER POTS, PLANTS, AND ENERGY

Seeds stored in a dry, cool environment can lie dormant for long periods of time. However, once seeds have moved from a state of dormancy to beginning growth, the process is irreversible and the application of light, heat, and water must be maintained for life to be sustained. Seeds and plants are both rugged and fragile. When improper amounts (too little or too much) of light, heat, and water are available, seeds and plants struggle to survive. When a point of "no return" is exceeded they begin to die and no amount of hastily applied light, heat, or water will resurrect them.

W = F x d

ENERGY = W = F x d

Seeds possess stored energy. When they are placed in the proper light, heat, and water they grow. This growth represents energy. Energy moves things. Energy is work. Work is done when a force (F) moves a body through a distance (d). Work is defined as force times distance.

Cut a two-liter pot. Nearly fill the pot with dry, lima beans. Insert a capless, two-liter funnel into the cylinder and push it down. Pour water into the funnel filling the lower portion up to the bean level. Place known weights inside the funnel. Observe the movement of the funnel.

***** Describe your observations.

Did you remember to count your beans?

How far did the expansion of the beans move the weighted funnel?

***** Devise a technique to measure the energy of one lima bean.

STACKING TWO-LITER PLASTIC CONTAINERS

Seeds can be planted in a single, two-liter container. With sufficient potting materials and water inserted through the top opening, seeds dropped into the container and positioned to your requirements (using a dowel or straightened wire, coat hanger) will grow. However, some care must be observed. Soil oversaturated in a low-evaporative environment remains too wet for too long a period of time for good seed growth to occur. Hence, the seed may mold and/or rot. This problem can be alleviated in several ways. One is to cut and remove the top portion of the container. This could be taped back in place at a later time. Keep the top off until you think the soil or other potting material is moist but not soaking wet, then plant your seeds. Keep the top portion off until the seeds have germinated and a plant has emerged. Another way is to punch holes in the bottom portion of the container. It is also recommended that you make a series of holes in the cylinder. These can be covered with small squares of scotch tape which can be removed as necessary to provide adequate ventilation. When a closed or nearly closed container is placed in direct sunlight, condensation takes place and a sweating action can be observed. You may wish to place the plants in indirect sunlight.

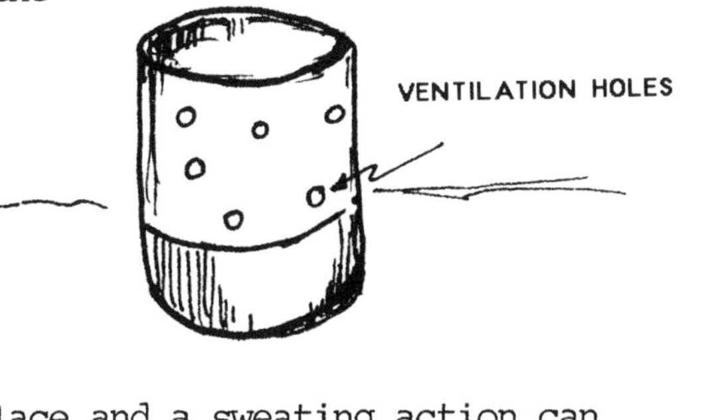

A two or more cylinder stack may be assembled. This is often necessary to accomodate plants whose height exceed the height of a single container. Stacking containers is a simple operation.

When cutting plastic containers to nest inside one another in a stack-like column, cut the upper container on its taper so that its diameter will be less than the full diameter of the container into which it will nest. A nesting done in this manner will require no taping and will permit you to separate these at some later date in order to water or on occasion aerate the seeds or plants.

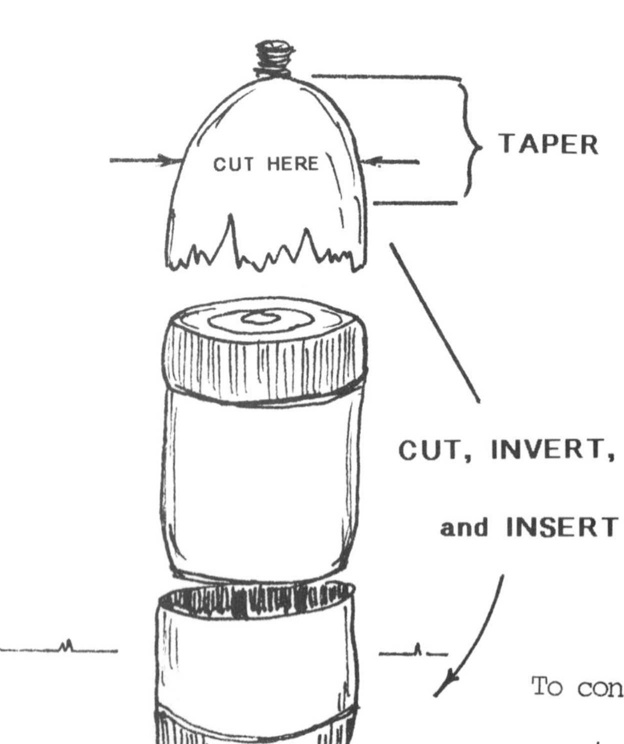

To construct the third stack and any subsequent stacks, remove the top portion and the bottom, black section of each container. Cut the remaining cylinder at one of the tapers and make the other cut a full-cylinder cut.

THE TWO-LITER HANGING POT

You can construct the two-liter hanging pot by first punching a hole in the cap. This is best done by inverting the cap and punching a hole from the inside out. This is a safety measure and it also minimizes cap breakage. Provide a cushion under the inverted cap to protect your punch's cutting edge as it penetrates through the cap. Tie a string around a small piece of wooden match. Glue the string to the match. Thread this through the cap. The match stick is retained inside the cap. Make a series of vertical cuts about one inch apart around the entire circumference of the cylinder. Carefully use a razor knife to make these initial punctures. In each puncture, insert scissors and cut parallel slits down the length of the plastic cylinder and completely around the cylinder. Squashing the container down, these vertical parallel strips will push out. When in a pushed-out, flattened manner, close to the crease staple each strip. As the plant grows, it will find its way out each opening. Remember do not over water your plant.

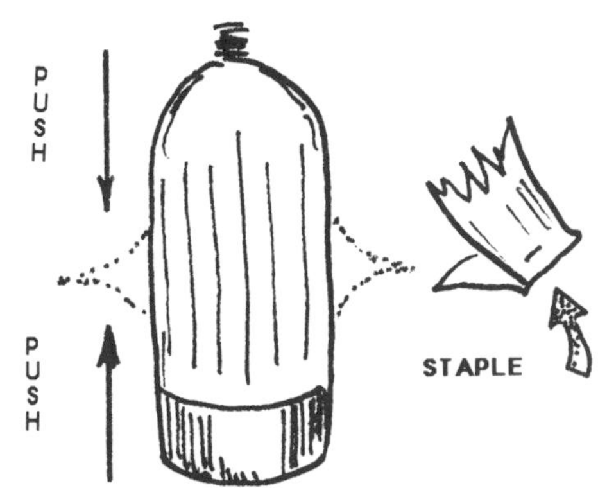

THE TWO-LITER AQUARIUM

Two-liter, plastic containers make excellent aquariums. The plastic container is unbreakable, thus removing concerns about broken glass and personal injuries. They are leak proof. With the cap screwed on, a closed system is established. If the system is a balanced one little care is required for maintenance.

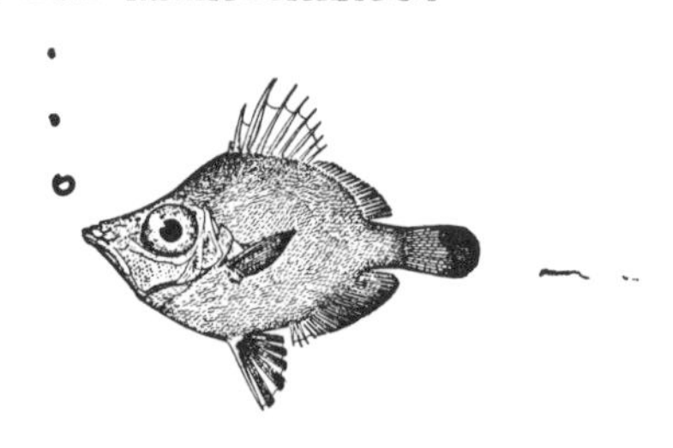

CLOSED

OPEN

An open aquarium is more easily prepared and subsequent adjustments can readily be made. Cleaning, etc. are more easily facilitated.

To construct an aquarium you will need sufficient gravel to anchor plants that you select. Tap water is usually adequate. Goldfish are easy to maintain and are usually recommended for your initial inauguration into the world of fish. For advice on exotic fish, plants, and gravel consult your local pet shop for first-hand advice.

QUESTIONS ABOUT PLANTS THAT FOSTER EXPERIMENTATION

***** How many plants can be grown in one, two-liter container?

- Is there a maximum number?
- What number is ideal for plant growth?

***** Is plant growth accelerated by using a plant food stimulant?

- How much plant food is too much?

***** Do plants really need soil to grow in?

- Can plant growth be facilitated only by water?
- Are there liquids other than water that can facilitate plant growth?

***** Does seed imbibing improve subsequent plant growth?

***** Is plant growth enhanced by the addition of vitamins to their liquid requirements?

***** Does talking to plants promote plant growth?

***** Will music stimulate plant growth?

- If so, which type of music?

***** Do plants react to an injury?

- Does this effect growth?

***** Do small lima bean seeds give rise to small plants?

- And conversely so, do large lima beans give rise to large lima beans?

***** Does the depth of seed planting effect plant growth?

***** What temperature is most advantageous for plant growth?

***** What humidity is most advantageous for plant growth?

***** What variety of water (tap, rain, or distilled water) is most beneficial to plant growth?

***** Would growing plants within an electric field be beneficial to plant growth?

***** What happens to a plant's root growth direction when the planting pot is inverted after initial plant root growth has started?

Select one or more questions and restate these questions as hypotheses suitable for experimentation.

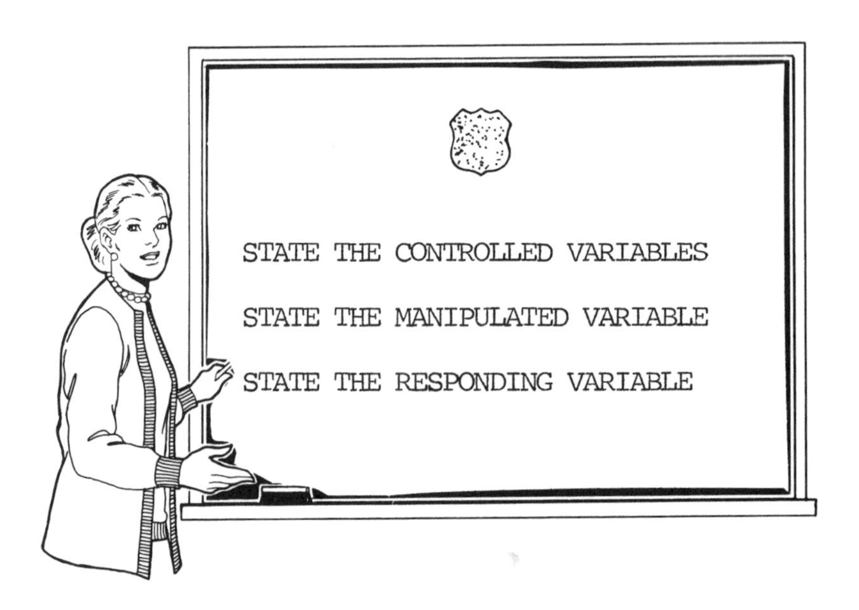

Take the necessary action to support or refute your stated hypothesis.

THE TWO-LITER PILL BUG CONTAINER

Pill bugs or land isopods may be found under stones, boards, logs, and other dark places. Pill bugs can be a useful addition to science instruction. A study of their behaviors as to movement, preferences for specific environments, etc. can provide interesting areas of investigation.

Their usual ecological environment can be readily duplicated in the classroom in a two-liter, plastic container containing damp, rich humus and small rocks or semi-decayed logs under which the organism can hide.

Pill bugs are hardy and multiply quickly in culture. Isopods relish bits of ripe fruit (apples), bits of lettuce, and occasionally pieces of raw potatoes.

LAND ISOPOD (PILL BUG)

To construct your pill bug container, cut out a sizeable window from a two-liter container. Separate the black sections from two other containers. Cut these so that they cradle your horizontal pill bug container.

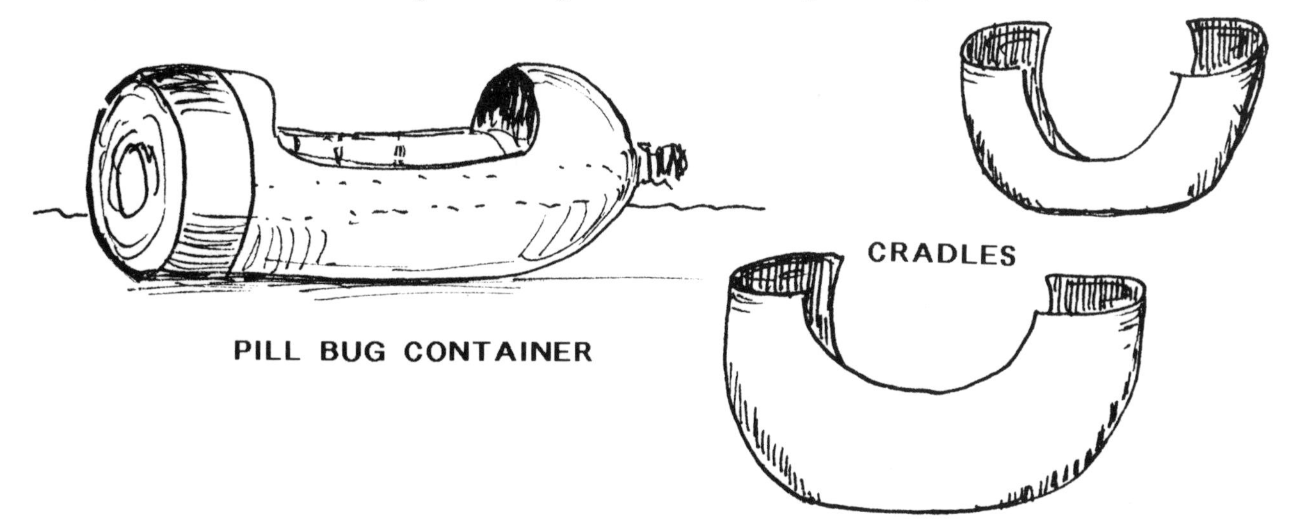

From another plastic container cut out another window much like the one you cut from the first container. Cut this window so that all its dimensions exceed those of the first window by one or two inches all around. This larger window serves as a cover over the first window area. Rubberbands will hold it tightly in place.

THE TWO-LITER TENEBRIO BEETLE (MEALWORM) CONTAINER

Tenebrio molitor (mealworms) are bred as food for laboratory fish, amphia, and some reptiles. One can purchase a beginning culture from local pet shops. The culture is then placed into your two-liter container with a sufficient quantity of slightly, moistened wholewheat flour, oatmeal, cornmeal, or bran. There should be sufficient material to cover the mealworms. An additional depth of two to three inches of food material (for example, oatmeal) should be provided. This will allow the mealworms to move freely about within the material. They will find their own way through the material and select an area best suited to them. Occasionally, spray a small amount of water into the container to keep the food material moistened. An occasional apple or potato slice added to the container would be a nice, mealworm treat.

To construct your two-liter, mealworm container, cut the bottom portion from a container. Invert the funnel top (with the cap intact) so that it can be inserted into the upright, bottom portion of the container. Pierce the funnel with numerous holes to provide some ventilation to the system. Keep the funnel in place so that flying adult beetles cannot escape. The funnel, which fits snugly inside the bottom portion, can easily be removed to add food, moisture, or remove mealworms.

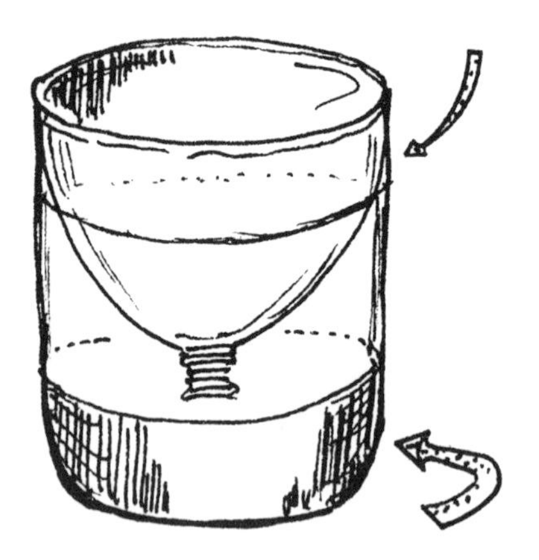

New colonies can be started by placing pupae in fresh food medium. At a temperature around 30°C (86°F), the complete life cycle may take four to six months. With new cultures, add shredded carrots placed over the top surface once a week.

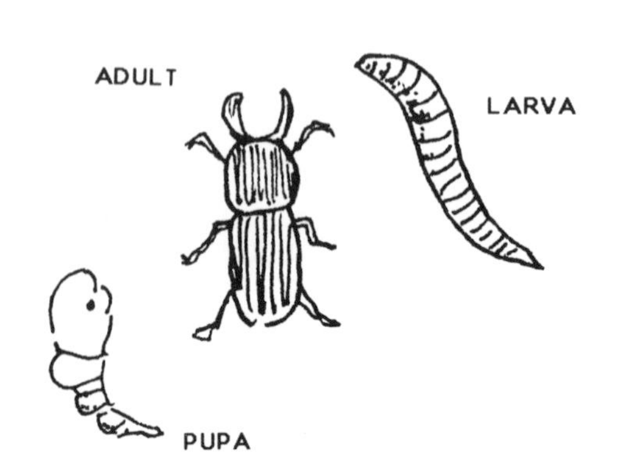

THE TWO-LITER TERRARIUM

A two-liter, plastic container is a natural for a terrarium. This can be expanded to a two or more stacked terrarium (see stacking two-liter containers). Terrariums are usually closed systems in which conditions approach a balance or a near balance one component with another. This balance or equilibrium within a closed system is not a speedy accomplishment. Plants, insects, or whatever else are introduced to the terrarium, each must find its own niche within the closed environment. Plus, they must adjust to the temperature and light controls that you provide. In a sense, you create a world of your own choosing. Whether it survives or not depends on many variables. Your observations of the terrarium will tell you what the evolving health of your terrarium is and is becoming. Adjustments must be made in response to your observations...more water/less water, more light/less light, more plants/less plants, more...less... and so it goes. The terrarium educates the viewer as to its needs. It tells you what it needs, if you read the signs. Initially one must fuss with the variables until the terrarium has stabilized. The results are worth the effort.

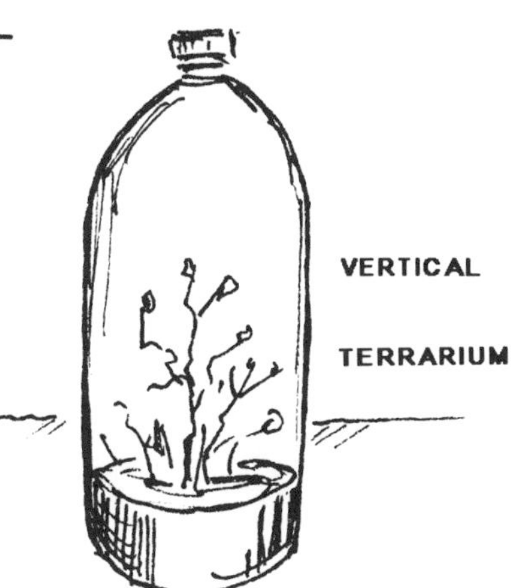

Plant your plants using a straightened wire, clothes hanger, a dowel, or any other device that does the job. Add soil and water. Add any insects you wish, for example, ants, flies, ladybugs, etc. Observe your terrarium and if necessary make any adjustments to bring your terrarium into a self-maintaining balance.

Separate the black, bottom sections from two other containers. Cut these bottom sections to cradle your horizontal terrarium.

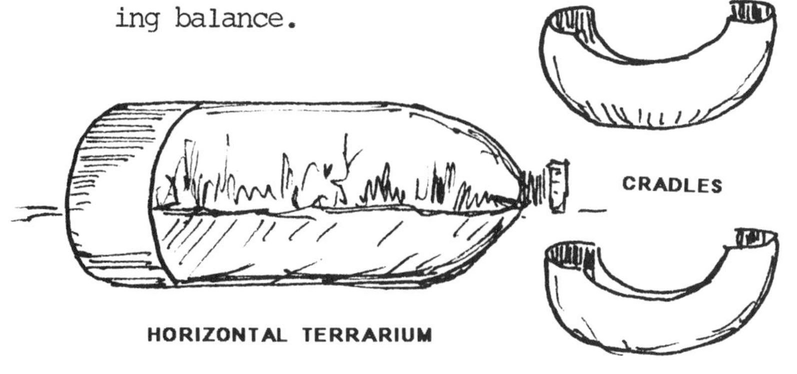

CRADLES

TIERED, TWO-LITER POTS

Two-liter pots can nest together in a tiered column. Any number of two-liter pots can be assembled to form a pot column. To construct your tiered,

two-liter pot column, cut out a bottom section from a two-liter container. Reserve the top, funnel-shaped portion with the cap intact. Cut several containers in the same manner, again reserving the top portion. Into one bottom section of the plastic container, place one, inverted, two-liter funnel top with cap intact. Push this down so that it fits snugly into the cylinder. Repeat this same operation inserting another two-liter funnel top into the first, funnel top. This action can be repeated until you have the desired number of pots in place.

If you desire each pot to be a closed system, the highest placed funnel top cannot be used as a valid, growing pot area. Each pot below this one forms a plug sealing the container below it. If you do not wish to use closed containers, punch holes in each lower pot's perimeter to encourage ventilation. There is no limit to the number of ventilation holes that can be formed. In any event, they are easily sealed off by placing a scotch-tape patch on any holes that you do not wish to use.

Seeds can be planted in each chamber. The funnel sections can easily be separated one from another to accomodate watering and subsequent plant attention.

THE TWO-LITER MOBILE

The two-liter mobile appears to be an innocuous toy. It is, but it can be elevated to where it can serve science instruction extremely well. To construct a two-liter mobile, you will need to make a hole (approximately 1/4 of an inch) in the center of the container bottom and the container cap. A long rubberband (or several smaller rubberbands knotted together) whose length does not exceed the height of the plastic container should be threaded through the container from the bottom to the cap. This would be a difficult task unless you fashioned from thin, stiff wire a hook, pull-through device. A wire clothes hanger can be fashioned into just such a device.

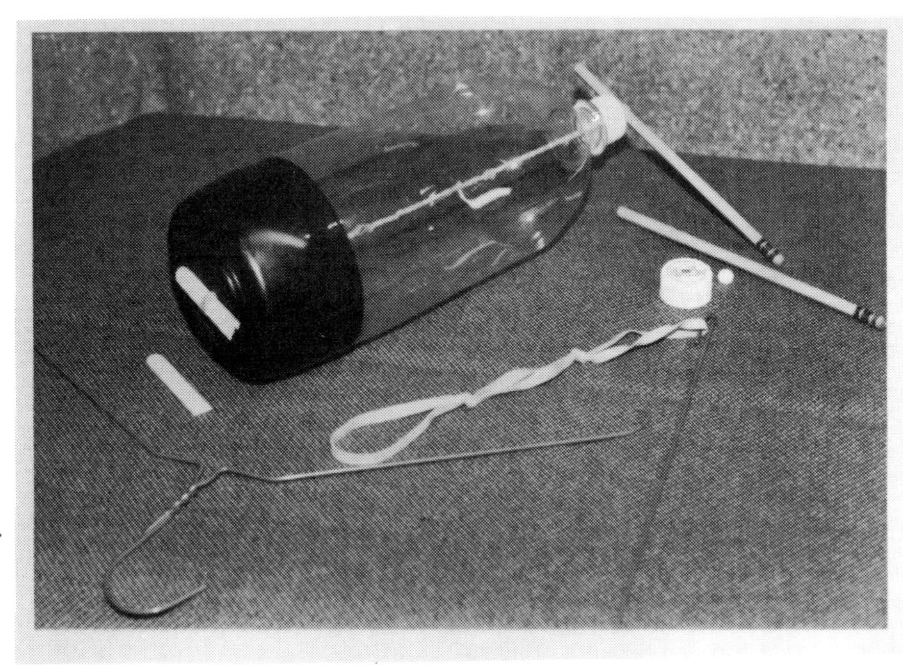

Start by placing one end of the rubberband on the hook of your wire device. Pull the rubberband into the container. As the rubberband is nearly, completely drawn into the container, insert an object (match stick, paper clip, section of a straw -- none of whose length exceeds the diameter of the container) inside the loop of the rubberband so that it is not pulled inside the container. Maintain tension on the rubberband so that the inserted object does not fall out. Thread your hooked wire and rubberband through the container cap and through a small, plastic bead. Beads, of all sorts, are readily available at craft supply houses. Through the rubberband loop emerging out from the bead, insert a pencil (straw, dowel, etc.). Screw the cap onto the container. The rubberband should be tight enough to hold both inserts (one at each end) under sufficient tension so as to not slip out of position. The two-liter mobile is now operational. Windup the mobile at the cap end. Check it to make sure that the insert at the opposite end remains stationary while winding up the rubberband. If it moves you must hold it, otherwise winding the rubberband becomes a self-defeating exercise. Place it on the floor and let it go.

SOME THINGS THAT MIGHT EFFECT MOBILE PERFORMANCE ARE:

- The rubberband is too long. Hence, when winding it up even after numerous cranks, the rubberband is still limp devoid of sufficient tension to move the mobile.
- The rubberband is overly tight and winding it puts too much tension and energy into the system.
- Too much initial winding of the mobile introduces too much energy into the system. Thus, when the mobile is placed on the ground it does not move in an orderly, straight-line direction. Instead, it spins erratically in place. Sometimes it even rises on end expending excess energy as it attempts to reach a state of equilibrium.
- The plastic bead in contact between the pencil and the container cap reduces friction between these two items. The pencil or straw dragging across the face of the cap increases friction and retards forward motion.

EXTENSIONS...

***** What is the maximum number of rubberband cranks for the maximum distance traveled by the mobile?

***** How does doubling the rubberband effect speed, direction, or distance traveled by the mobile?

***** How does increasing the mass of the mobile effect the speed, direction, or distance traveled by the mobile?

***** How does increasing the mass of the mobile effect the speed, direction, or distance traveled by the mobile when the number of cranks of the mobile are kept constant?

***** In terms of performance as to speed, direction, or distance, what is the best position for the pencil? Should it be positioned so that one-third of the pencil extends beyond the rubberband and two-thirds of the pencil drags as a rudder? Or does a mid-point positioned pencil work best?

***** Numerous other variables can be investigated, for example, the type and variety of bead, the lubrication of the rubberband (use of talcum powder), etc.

PICK AN AREA OF INVESTIGATION

ACTION

- Formulate a hypothesis
- Identify the controlled, manipulated, and response variables
- Conduct the experiment

The two-liter mobile moves. Things that move, move through a space and this movement takes time. This process is called a space/time relationship. Space/time relationships can be plotted on a graph. A graph is a device for communicating a relationship between two variables. In this instance time versus distance (spatial) is shown.

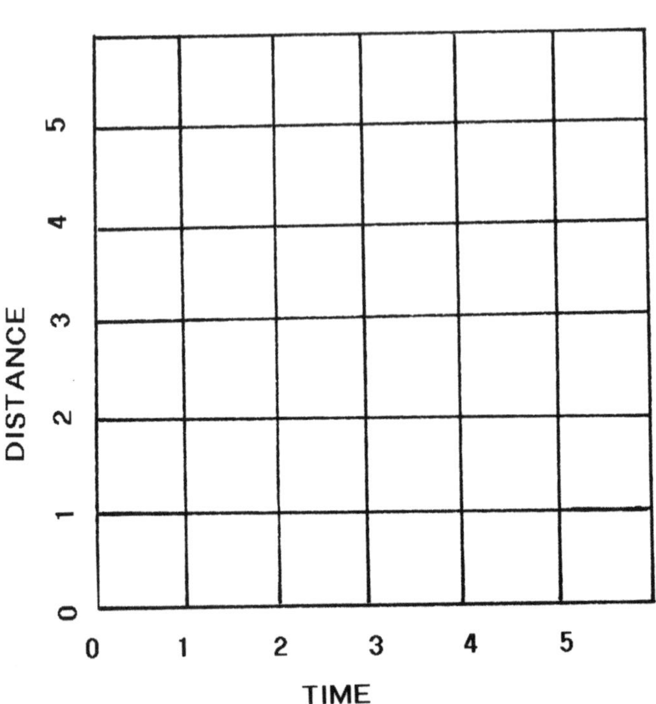

A graph is a source of much information. Graphs provide data in a concise, pictorial form. From this data interpolations and extrapolations about data can be made. The action of the two-liter mobile in response to the large variety of manipulated variables that exist affords numerous opportunities for graphing and the interpretations that arise from observations of the accumulated data.

THE TWO-LITER BALANCE

Every science classroom should have one or more balances readily accessible by students. Every student should make a balance. One that is peculiar or unique to the student. One that is operable by its creator and which becomes a piece of equipment that permits the student to obtain quantifiable data about an event and to observe data changes about that event over periods of time. The best balance in the hands of an uninformed novice is reduced in efficiency by the level of the investigator. An adequate balance in the hands of someone who understands a balance and how it operates can become an excellent piece of equipment for quantification.

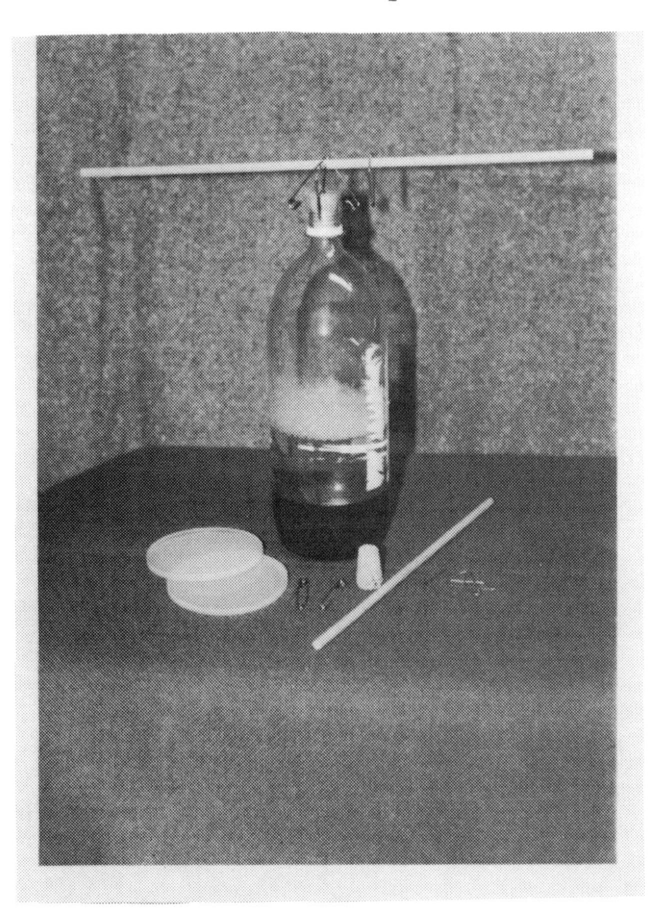

The two-liter balance can be constructed so as to furnish fairly sophisticated measures. To construct a two-liter balance, you will need a two-liter container, a 9/32nd or 5/16th of an inch diameter wooden dowel (approximately eighteen inches in length), a cork that fits the two-liter opening, two large safety pins, two oleo tub covers, string or fishing line, and one, large paper clip.

To construct your two-liter balance, add approximately one liter of water to your container. The water serves as ballast and keeps the container upright and steady. Find the center of your dowel. Drill a small hole through the dowel. Drill two additional holes through the dowel. Each should be drilled in one inch from the ends of the dowel. And, they should be drilled ninety degrees out of phase with the first drilled hole.

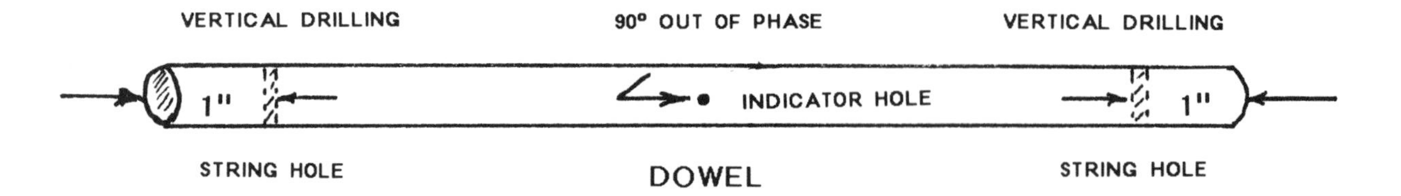

The two end holes must be large enough for string or fishing line material to pass through them several times. The mid-point hole is for the paper-clip indicator which must pass through the dowel.

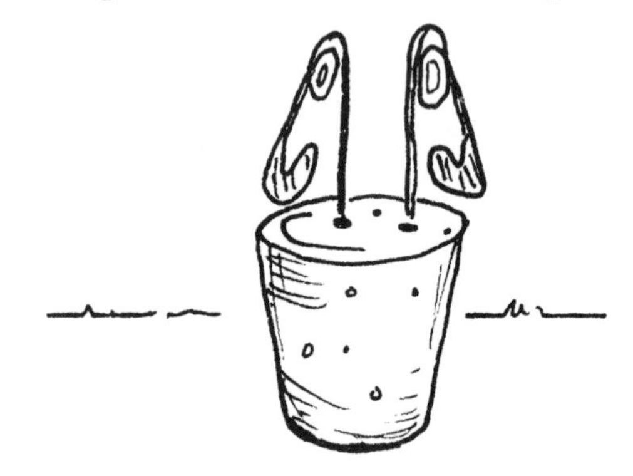

The corking apparatus consists of two, large safety pins. These are pushed into the cork far enough apart for the dowel to be inserted between them. It works best if the pins face away from one another.

Straighten out one large paper clip to form the balance-movement indicator.

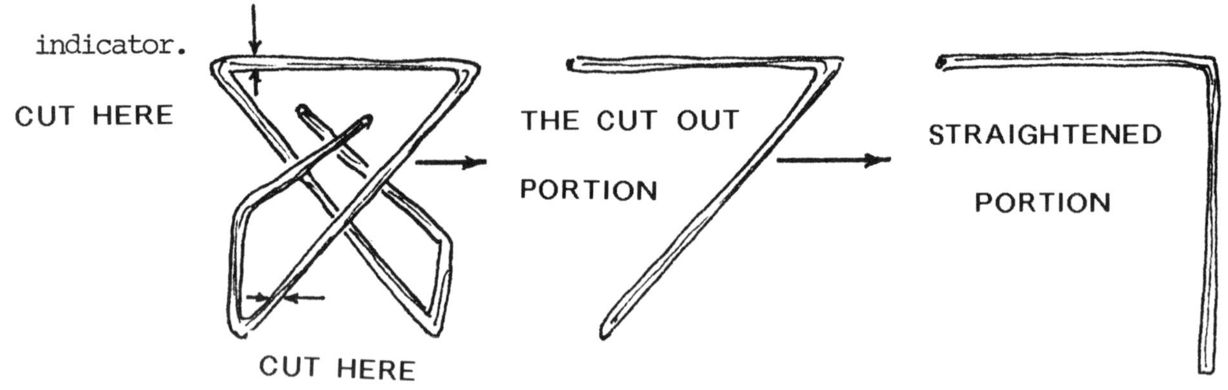

Insert the straightened paper-clip, indicator by threading it through one safety pin, through the inserted dowel, and then through the remaining safety pin. The paper-clip indicator not only serves as an axle for the swinging dowel, but the front, pendant portion becomes an indicator of the movement of the dowel. The dowel and the indicator should swing as one. If they do not, glue the paper-clip indicator at the points of entry and exit from the dowel. Do not glue the indicator to the safety pins.

Punch holes in the oleo covers. These holes are necessary to accomodate strings that join the covers to the dowel. Arrange these covers so they hang approximately the same distance down from the dowel.

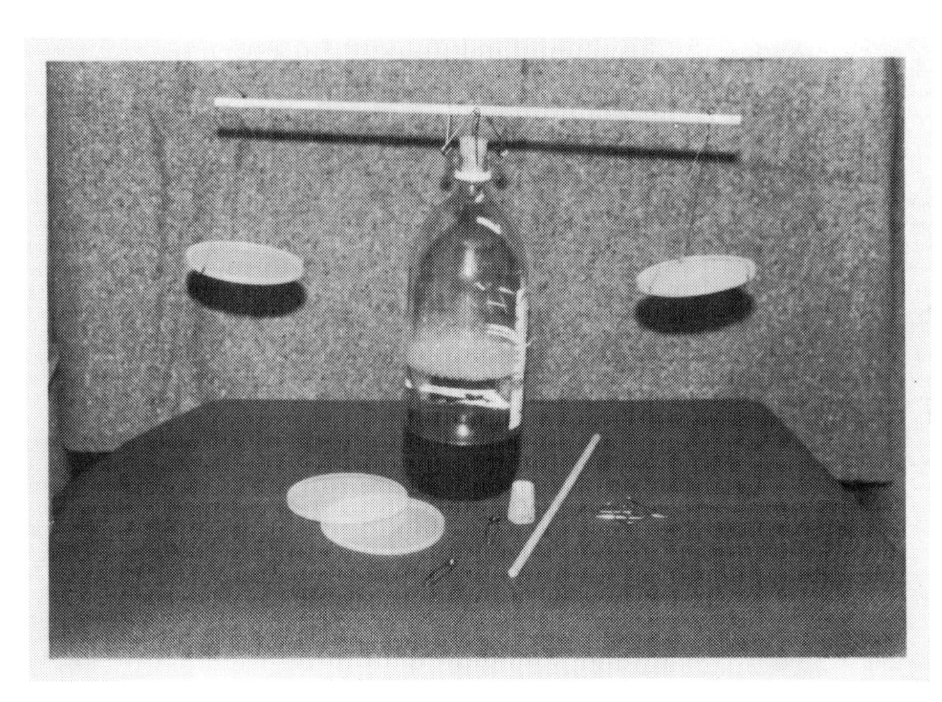

Most likely your balance will not be in balance. This is correctable by adding a weight in the form of a rider to one side of the balance. Place the rider on the arm of the balance that is elevated. Move the rider to bring the balance into equilibrium. At this point you may need to snug the rider tightly to the dowel to avoid slippage resulting in inaccurate measures.

REFINING THE TWO-LITER BALANCE

Almost any item may be used in the determination of the mass of another object. One could use match sticks, thumbtacks, paper clips, or BBs, and the mass of the object would be expressed in the unit utilized, for example, an object could weigh 19 paper clips. This information is more useful if it is expressed in known units like grams. Thus, it is recommended that one establish a prior, calculated conversion factor, for example, six gigets equal one gram, ten wifs equal one gram, and so forth.

Weights recorded in whole numbers may not be satisfactory for your needs. The two-liter balance can be adjusted to measure smaller units such as one tenth of a gram or even one twentieth of a gram. This refinement can be achieved by balancing the balance and then placing a known weight (one gram) in the pan opposite the side with the rider on the dowel. With the balance now out of balance, mark a spot on the dowel at the current position of the rider. Move the rider out and bring the balance back into balance. No weight has been added to the balance and yet balance was achieved. This was done by moving the rider through a specific distance on the dowel arm. With the balance in balance, make another mark on the dowel at that point where the rider finally was positioned. This distance between the two marks on the dowel should be divided into ten equal spaces. There is a simple technique to perform this task despite the length of this space. Create a small paper ruler that divides this space into ten equal units (see "How to divide any line segment..."). Return the rider back to its original point.

Remove the one gram weight from the pan. Any subsequent item being weighed would be placed in this pan. Weights would be added to the pan on the rider side of the balance until balance is achieved. Usually the last weight added is too heavy and the previous one is too light to bring the balance into balance. Hence, a portion of a unit, say one tenth of a gram, is needed. With the observed weight being shy of the last unit of weight, hold your paper ruler up to the rider arm (between the two, previously established, marks) and while doing this move the rider along until the balance is balanced. Now read on your paper scale what fraction of the distance is shown and mentally add this conversion in grams to the sum of the weights showing in the pan. The final weight should reflect whole numbers and a fraction of a whole number. Your balance has been refined to provide you with one tenth of a gram measure.

HOW TO DIVIDE ANY LINE SEGMENT INTO ANY FIXED NUMBER OF PARTS

Teachers find this technique extremely useful. Teachers, on occasion, have requirements for establishing equal columns within a fixed space, for example, dividing an 8 1/2 inch width of paper into a set number of equally spaced columns. This can be accomplished by using the top of the paper as a line (AB). Let the left margin of the paper be line AC. If one wished to divide the 8 1/2 inch width sheet into nine evenly spaced columns, one would place a ruler so that nine, equal divisions are included between point B and a point on the perpendicular line AC.

The same procedure can be used to divide any line segment into any fixed number of parts.

For example, to divide line segment AB (or any other length of line) into ten divisions, you would place the zero point of the ruler on line AC so that ten equal divisions are included between point B and a point on the perpendicular line AC. Half measures or any other consistent increment on the ruler can be used for subsequent equal divisions.

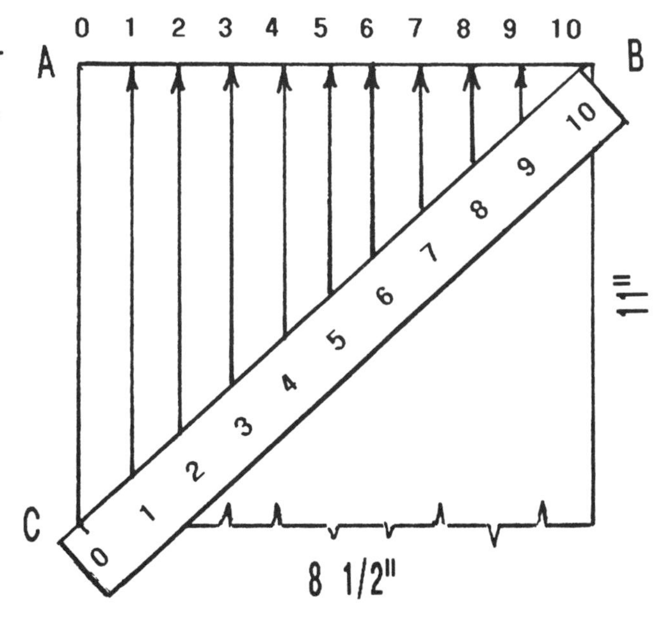

This technique can be used to divide your paper ruler, noting the distance between the two rider marks on your balance, into ten equal parts.

THE TWO-LITER RAISIN RISER

Uncap any two-liter, carbonated beverage container. A clear beverage and a clear container are preferred. Quickly place four or five individual raisins in the container. The raisins will sink to the bottom of the container. They will remain there for a short period of time, and then rise to the surface.

While floating at the surface of the beverage, they may gyrate a bit as gas bubbles are released and then sink to the bottom to repeat this cycle.

Carbonated beverages contain the gas carbon dioxide. The raisins in the container collect small bubbles of carbon dioxide gas. These bubbles buoy up the raisins and they rise to the surface. Here, they lose some or all of the carbon dioxide bubbles to the air. The raisins having lost some or all of their buoyancy are too heavy to float and thus sink to the bottom only to collect more carbon dioxide bubbles. The raisins are again buoyed up to the surface. And, the cycle repeats itself until the carbon dioxide gas contained in the beverage is spent.

I WANNANO:

***** Which commercially available beverage recycles raisins for the longest period of time?

***** Which variety of raisin (yellow versus black) recycles for the longest period of time?

***** Does the mass of individual raisins effect the rate of recycling?

***** Does the mass of individual raisins effect the rate of ascent and/or descent?

***** When raisins are clumped in groups of two, three, or more raisins, will the carbon dioxide gas bubbles be able to raise these clumps to the surface?

***** What happens to the total exposed surface area of clumped raisins as compared to the total exposed surface area of the individual raisins? How does this effect buoyancy?

***** Can carbon dioxide buoyancy work with the raising of other objects?

***** Does the texture of the surface area of objects effect carbon dioxide buoyancy?

***** How high in a column of a carbonated beverage can raisins rise?

***** Will other dried fruits or berries rise when placed in a carbonated beverage?

FACT: Small prunes when placed in a carbonated beverage may float. They may float for a period of time until they imbibe a sufficient amount of water to cause them to sink. For a period of time sufficient gas bubbles accumulate on them and this may recycle the prunes. However, as the prunes continue to take in water, the mass of the prunes soon exceeds the buoyancy power of the collected carbon dioxide gas bubbles and the prunes remain on the bottom of the container.
In this activity it is sometimes necessary to track one specific raisin or prune. A small drop of colored nail polish placed on a raisin or prune enables one to follow the path of a specific raisin or prune.

Select one or more questions and restate those questions as hypotheses suitable for experimentation.

- STATE THE CONTROLLED VARIABLES
- STATE THE MANIPULATED VARIABLE
- STATE THE RESPONDING VARIABLE

Then take the necessary action to support or refute your stated hypothesis.

THE ONE OR TWO, TWO-LITER WATER CLOCK

You will need two or more two-liter plastic containers to construct your water clock. You will need a ring stand to hold one, two-liter container and you will need a plastic, catch tub.

Place an empty, capless two-liter container inside a catch tub. Pierce a two-liter container cap making a small hole in its center. Fill a two-liter container with water. Cap it. Capping the hole in the cap, invert and suspend this container directly over the empty, two-liter container. Make any adjustments necessary to make sure the water drops into the bottom container. Time the interval it takes for two liters of water to be emptied into the bottom container. An alternative to the two-liter water clock is the one-liter open container water clock.

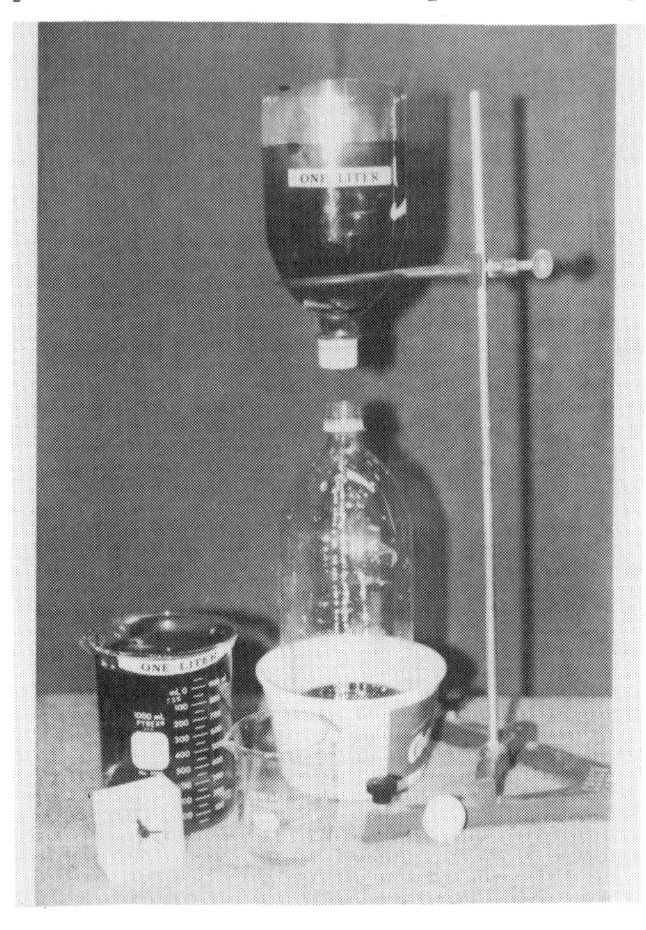

If it takes two hours for two liters of water to drip out then the rate would be one liter, per hour or one-half liter, per half hour. This water clock would be described as a liter-per-hour clock. If it takes one hour for two liters of water to drop down, your water clock would be described as a two-liter-per-hour water clock. Describe your water clock's performance.

I WANNANO:

***** If the size of the cap opening was enlarged (or reduced), what effect would this have on the time it takes to empty the two-liter container?

***** If the density (and/or viscosity of the liquid) is changed what effect would this have on the time it takes to empty the two-liter container?

***** If the temperature of the liquid is increased (or lowered), what effect would this have on the time it takes to empty the two-liter container?

THE TWO-LITER SPATIAL EXERCISER

Direction, position, and spatial considerations are all part of communicating in science. Experiences can be provided for children to state where they are, where they have been, and where they are going. Early spatial exercise instruction with children may be confined to above/below, in front of/ in back of, and to the right of/to the left of. The two-liter spatial exerciser can advance this learning.

Using any number of two-liter containers, arrange them in any configuration of your choosing, for example, a cross. Individuals select a container. As the instructor, you also select a container. And, you place your container anywhere within the configuration. The children select their location. Each child then states where he/she is in relation to the teacher, for example, "I am in front of the teacher." Of course, descriptions will change as you change your position within the pattern. At some point, you may wish to elevate or raise your container in the air so that a new dimension can be added to student responses.

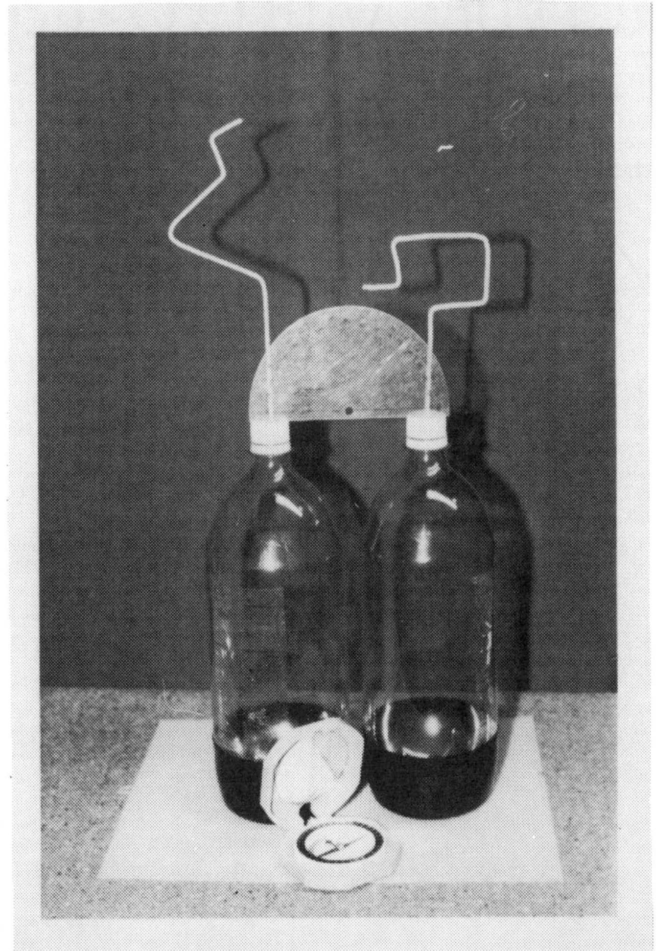

Additional experiences can be provided using the two-liter spatial exerciser. To construct a two-liter spatial exerciser, make a small hole in the container cap. Place a pipe cleaner in the opening.

Bend the pipe cleaner to show only right angle turns. This can be expanded to angular turns. The child assumes the role of an ant climbing up the pipe cleaner. As the ant progresses up the pipe cleaner, the ant communicates the various directions he takes in his travels.

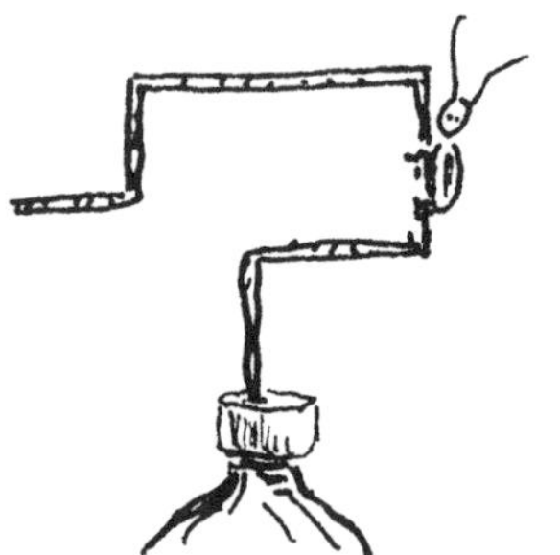

As a natural progression you can repeat this incorporating a variety of angles such as right, obtuse, and acute angles. Upon mastery of this, include a ruler, a protractor, and a compass so that more refined statements about direction and position can be made. Add ballast (sand or water) to your container to provide for greater container stability. When including a compass for children to use, place a mark somewhere on the container that can be aligned with north. Thus, the compass and the container can be synchronized for more accurate descriptions of direction.

BOWLING WITH TWO-LITER TEN PINS

Fill ten, two-liter plastic containers about 1/3rd full with water for ballast.

Arrange these ten pins in a triangle much like ten pins in a bowling alley. Select a suitable ball, for example a six-inch, diameter rubber ball with sufficient weight to bowl over the ten pins. This should be played on a flat surface. It is recommended that two sets of ten pins be set up opposite from each other. This eliminates much ball chasing and helps in the resetting of pins. Assign a number value to each pin. The notion of spares and strikes can be included.

THE TWO-LITER OCEANOGRAPHIC DENSIOMETER

Hot water rises. Cold water sinks. Cold water is more dense than hot water. This phenomenon is one of the mechanisms that operates within oceans and lakes. This process can readily be shown utilizing two-liter containers.

To construct your two-liter oceanographic densiometer, you will need two, two-liter plastic containers, two container caps, and water. The success of this activity lies in the preparation of the container caps. From the inside out, punch (or drill) a 7/16 or 1/2 inch hole in each cap. Using one of the commercially available instant glues (such as cyanacrylate) and observing all the instructions and precautions printed on the label, glue the two caps together, back to back. Make sure the separate holes are aligned one with another. Following the directions on the glue container, let it dry thoroughly. Note: Not all glues are alike and not all work in the same manner. Therefore, some experimentation with different glues is advisable. If you do not feel comfortable that your glue will hold, wrap clear tape around the joint as a reinforcement.

Fill one, two-liter container with warm water (you may wish to add a few drops of food coloring to accentuate the visibility of water movement). Fill the remaining two-liter container with cold, clear water. Screw the double cap on this container. Invert the container and screw the cap into the bottom container. Initially there will be a small amount of leakage. The warm, colored water on the bottom will rise up into the upper container. The warm water replaces the cold, clear water in the upper chamber. The cold water sinks to the bottom of the lower chamber. Reverse the process...warm colored water in the top chamber and cold water in the lower chamber. Describe your observations.

THE TWO-LITER CONTAINER TORNADO REPLICATOR

To construct your two-liter container, tornado replicator follow the same steps as outlined for the oceanographic densiometer. The same apparatus is used, however, the amount and placement of water differs. Fill one, two-liter container with water approximately two-thirds full. You may wish to add a few drops of food coloring to accentuate the visibility of water movement. Screw the double cap on this container. Invert the entire, two, two-liter container apparatus. With the bottom container resting securely on a flat surface, using both hands, rotate the top container in a circular manner.

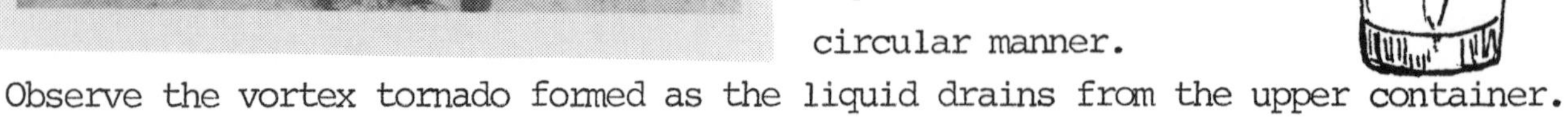

Observe the vortex tornado formed as the liquid drains from the upper container.

THE TWO-LITER BUBBLE-ATOR

The two-liter bubble-ator is a captivating view of bubbles. Locked within a closed container these bubbles retain their structures for one-half hour to one hour.

To construct your two-liter bubble-ator, take a clean, empty, two-liter container, add a small amount of Joy, clean rinsing, dishwashing liquid, fill the container with water, completely drain the water, and then cap the container. With some shaking, numerous lattice-like structures will appear. Some are three-dimensional structures exhibiting irridescent planes. Some are pentagonal honeycombs. And others form a multitude of small, multi-sized circles and spheres resembling a milky way, stellular mass.

The bubble-ator's display disappears after a one-half to one hour period of time. However, the bubble display can be rejuvenated by re-shaking the container and its contents.

The bubble-ator is enhanced by viewing the bubbles in sunlight or by viewing them in darkness and rotating the container slowly in front of a flashlight.

I WANNANO:

***** What is the most common geometric shaped displayed within the bubble-ator?

***** How does the quantity of soap effect the number and variety of bubbles?

***** What materials could be added to the soap, for example, food coloring, glycerin, sugar and what would their effect be on the number, structure, or size of the bubbles?

THE TWO-LITER BAROMETER

Air pressure is measured by a device called a barometer. A barometer reacts to changes in air pressure. To construct your two-liter barometer, you will need two, two-liter plastic containers, a ruler, and water. Remove the top from one, two-liter container. Discard the top, funnel portion. The remaining cylinder should be about seven inches in height. Fill the other two-liter container with water tinted with a few drops of food coloring to enhance visibility. With the cap off, slip the seven-inch high cylinder snugly down over the top of the water-filled container. Invert the entire apparatus. Some water will run out of the upper chamber and into the lower chamber. With the two-liter barometer in this position, make several triangular cuts, placed as high above the water line as possible. This opens the system to outside air pressure. Attach a ruler to the sidewall of the container. Record the present reading of the water level.

Air pressure presses down on the surface of the water. If the air pressure rises, more water is pushed up into the upper chamber. This is reflected in a slightly higher scale reading. When air pressure fails, water inside the upper chamber has less resistance to overcome in its attempt to seek its lowest level, hence a small amount of water comes out and a slightly decreased scale reading is observed.

TRIANGULAR CUT TRIANGULAR CUT

THE TWO-LITER AIR CONVERGER

To build your two-liter air converger, you will need one, two-liter container, a birthday candle, a clothespin to hold the birthday candle erect, matches, and a small dish or aluminum foil to catch the melting wax.

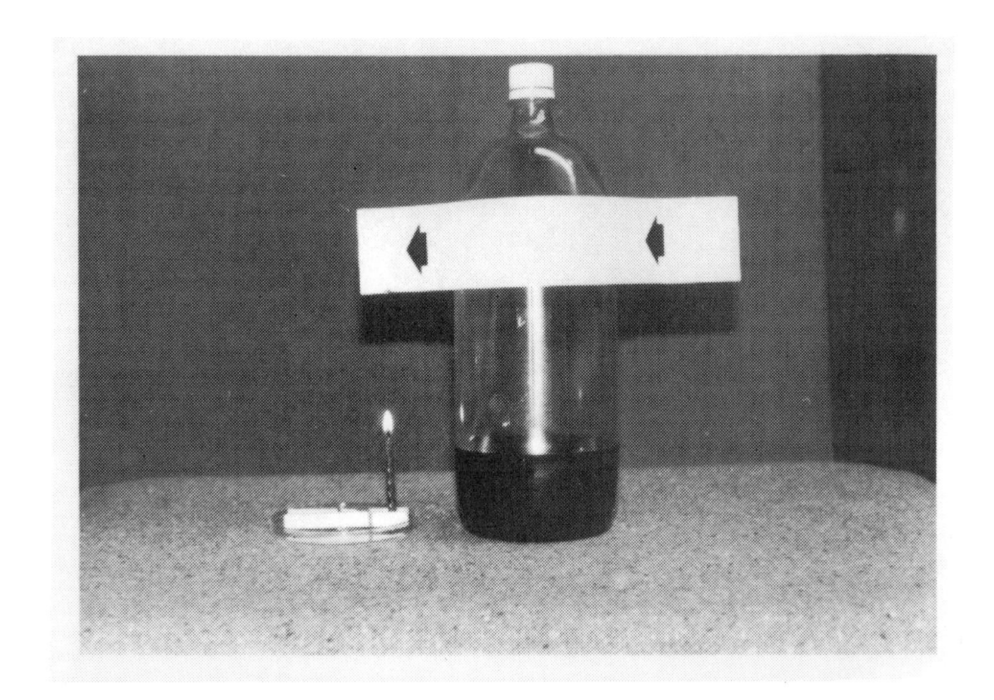

For ballast, add water to the container. Place the candle behind the two-liter container. Light the candle. From a position directly behind (and opposite) the candle, blow (in the direction of the arrows) towards the burning candle. The flame, hidden behind the container, will be blown out because the stream of air strikes the two-liter container and follows the contour of the container converging on the other side.

I WANNANO:

***** What is the maximum and minimum distance the candle must be placed in order to be blown out?

***** What would be the maximum or minimum distance the person doing the blowing must be positioned in order to blow out the candle?

***** What effect does the blowing have on the direction of the candle flame?

THE TWO-LITER AIRGUN

The two-liter, plastic container airgun is an activity that can enhance science instruction in numerous process skill areas, such as, observing, measuring, graphing, and interpreting data.

To build your two-liter airgun, you will need one, two-liter container, a birthday candle, one clothespin, a piece of thin, rubber sheeting, matches and a small dish or aluminum foil to catch any melting wax. The two-liter airgun is constructed by pulling the black section off the container. Applying heat from a hair dryer to the bottom of the container while rotating the container will facilitate this separation. Once the bottom is removed, cut out the bottom of the container below its taper. Retention of a portion of the container taper is necessary so that the container can later be reinserted into the black section portion. From the black section portion, cut out a two and one-half inch diameter circle. This bottom piece fits back on the tapered end of the container. Before slipping this bottom piece back, place a piece of thin rubber or latex approximately seven inches square (this can be cut from large balloons or purchased from

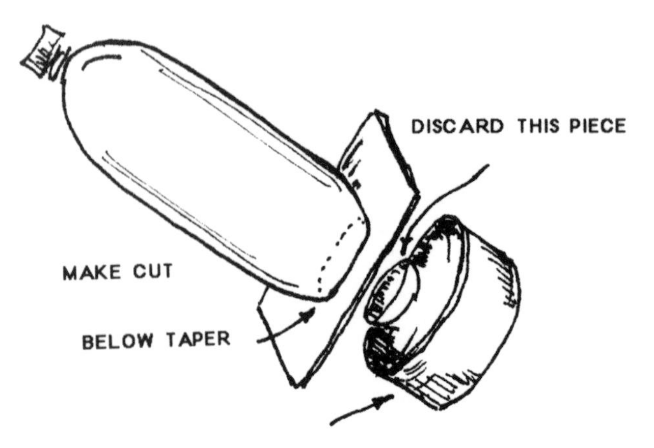

pharmaceutical stores) over the tapered end. This becomes locked in place when the black sleeve is reattached to the container.

The rubber sheeting acts like a drum head. It can be pinched with two fingers and pulled outward. This action fills the container with additional air. When this apparatus is pointed at the burning candle and the pinched rubber-sheeting is released, a ball of air strikes the candle and blows it out. Anchor your candle before attempting this activity. A burning candle is always dangerous and care must be taken to avoid a fire. Children should be instructed as to the inherent dangers of fire. If you have any doubt about the children's abilities to perform this activity safely, confine this activity to a demonstration by you.

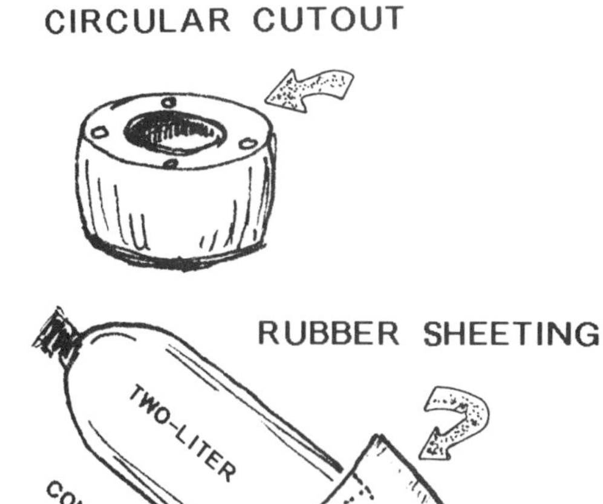

I WANNANO:

***** How far from the candle must the two-liter airgun be held to blow out the candle?

***** Is the air that is inside the two-liter container the same air that makes up the air ball that blows out the candle? How could this be tested?

***** How does the distance the sheeting is pulled out from the container effect the distance traveled by the air ball?

***** If the opening is restricted (screwing a cap on the container having pre-drilled a one quarter inch hole) how does this effect the distance traveled by an air ball shot from the airgun?

***** What is the speed of the air ball as it travels from the airgun to the target flame?

THE TWO-LITER VISCOMETER

The two-liter, container viscometer does not quantitatively measure viscosity. It compares viscosity. The viscometer serves to compare one, two, or more different liquids as to their viscosity. Thus, they can be rank ordered -- least viscous to most viscous. A liquid's resistance to flowing is its viscosity. In some liquids, the particles of material are attracted one to another. The more attraction the particles have, the less easily the liquid flows. The material tar does not flow readily. Its particles have a great

attraction one for another. By contrast water flows readily and its particles have less attraction for one another.

An object floating through a liquid is retarded more by liquids whose particles are strongly attracted to one another than in liquids whose particles are weakly attracted to one another. Heating liquids make them less viscous and objects flow more readily through them. Cooling liquids makes them more viscous and objects are retarded more when flowing through them.

Fill a two-liter container with water. Drop a bead or an imitation pearl into the water. Time the interval it takes for the object to sink to the bottom. Fill a second, two-liter container with cooking oil. Drop the same object into the cooking oil. Time the interval it takes for the object to sink to the bottom. How do these two, time intervals compare? **Which liquid** is the most viscous? Fill another container with **light-colored, motor oil.** Drop the same object into the motor oil. Time the interval it takes for the object to sink to the bottom. Rank order the three containers from the least viscous to the most viscous.

I WANNANO:

***** Is the rate of descent of an object dropped into a container filled with a specific liquid effected by the mass of the object dropped?

***** Is the rate of descent of an object dropped into cold water the same as one dropped into hot water?

***** If two objects of the same mass but of different shapes were dropped into similar liquids, would the rate of descent be the same?

PICK AN AREA OF INVESTIGATION

ACTION

- Formulate a hypothesis
- Identify the controlled, manipulated, and response variables
- Conduct the experiment

THE TWO-LITER MAGNIFIER AND REVERSER

A two-liter, plastic container filled with water can magnify objects viewed through it. The curved plastic container behaves like a convex lens. When convex lenses are brought very close to an object, it magnifies them, and the image remains in its original orientation. When light passes through a convex lens it converges to a point called a focal point. Light passes through this focal point and emerges continuing in a straight line. It is the passing through the focal point and continuing on that causes an image to be seen in an inverted form.

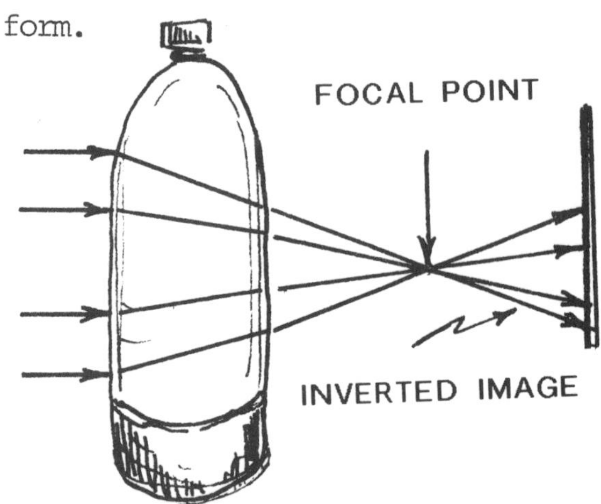

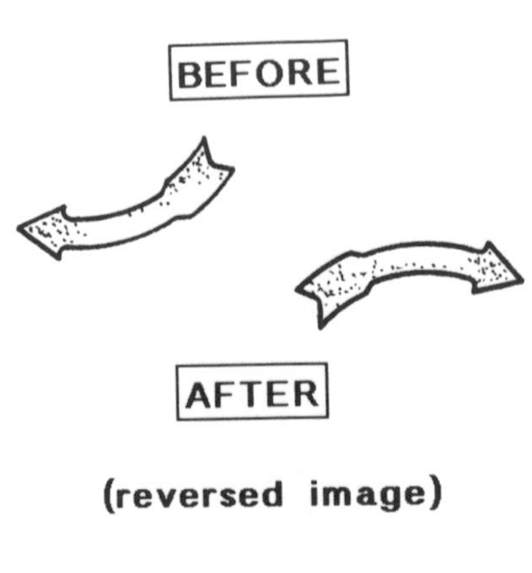

THE TWO-LITER CARTESIAN DIVER

You will need one, two-liter, plastic container. Fill the container to the top with water. Insert an eye dropper into the container. The eye dropper should be a little less than half-filled with water. The eye dropper should float in the container. If it does not, you will have to repeat the process. The amount of water inside the eye dropper is critical to this activity. Too much water and the diver will sink to the bottom and remain there. Too little water and the diver is so buoyant that you must use excessive force applied to the sides of the container to make the diver dive.

With the eye dropper floating at the top of the container, cap the container. Squeeze the sidewall of the container. What do you observe? Release the pressure. What do you observe?

When the jar is squeezed, water is forced up into the eye dropper compressing the air inside the dropper. With the added water, the dropper becomes less buoyant and it sinks to the bottom. When you release the external pressure on the container, the compressed air inside the dropper is free to expand and it forces water out of the dropper. The dropper now being less dense, floats back to the surface of the water. This action can be observed by watching the fluctuating water level inside the dropper in response to applied and released pressures on the container.

THE TWO-LITER ELECTROSCOPE

The electroscope is an electron detector. To construct your two-liter container electroscope, you must find a cork or rubber stopper that fits the container opening. Make a hole in the cork to accept a length of clothes-hanger wire. Bend this wire at both ends. The wire should be about one inch above the stopper. If the stopper or cork opening is too large, plug it with soft clay or putty. On the opposite end place a strip of aluminum foil. Insert the cork apparatus into the container.

Under normal circumstances, a substance will have the same number of electrons (-) and protons (+). This balance neutralizes or cancels out each other and no electrical charge is **apparent.**

When an object is rubbed, electrons are rubbed off. When a balloon is rubbed with nylon, wool, or fur, electrons (-) are rubbed off the fabric and onto the balloon. The balloon is negatively, electrically charged. To activate the two-liter electroscope, rub a balloon with nylon, wool, or fur. Touch the external, one-inch portion of the clothes-hanger wire. Electrons from the balloon will flow into the clothes hanger wire and down into the aluminum foil. The separate pendant portions of the foil will become negatively charged and will separate because they repel each other. If a positively charged object is brought near the wire head, the aluminum pendant portions will come together.

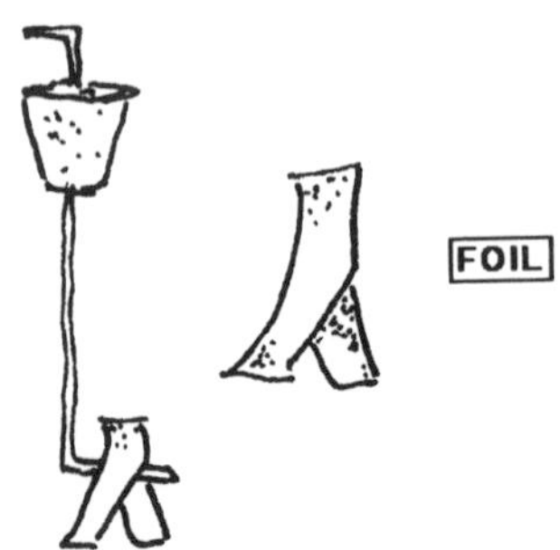

THE TWO-LITER COLLAPSING CONTAINER

Air exerts pressure. The pressure of air at sea level is approximately fifteen pounds on each square inch of surface. Differences in air pressure tend to equalize, one with the other. A heated gas expands and this expansion results in less air within a given volume.

To demonstrate the collapsing container, use two, two-liter, plastic containers. One will be a capped container filled with air. The other container is filled with warm water. The warmer the water, the better results. Let the container with warm water in it stand for a minute or two. Pour out the warm water. Quickly cap the container. Run cold, tap water over each container. What do you observe?

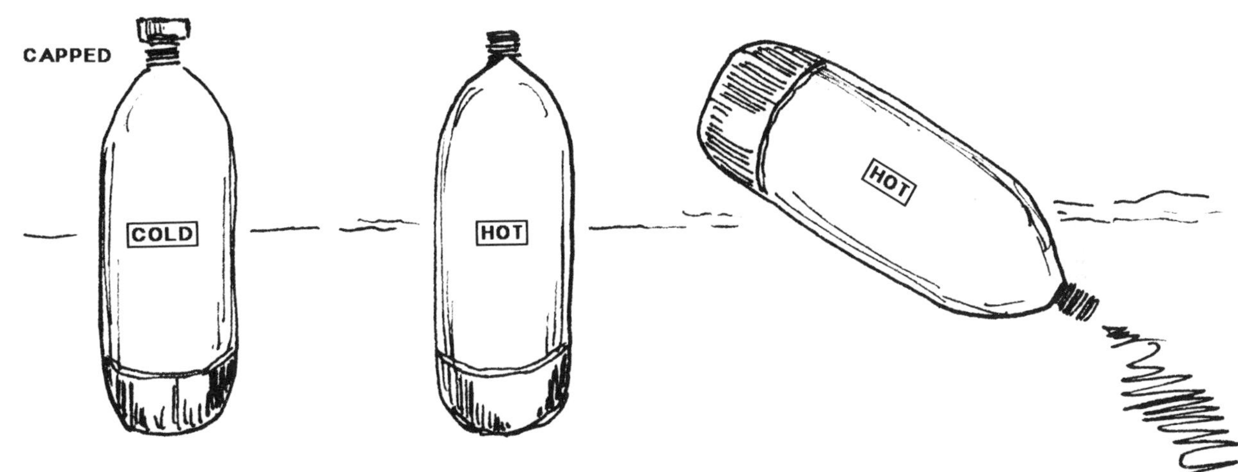

An alternative to filling one container with warm water is to place an uncapped container in a pan of warm water holding it upright. Continue this for a period of two to three minutes. Quickly cap the container. Then cool it down.

The application of warm water and its subsequent removal followed by a quick capping of the container results in less air in the container and, therefore, less pressure. As a result, the greater, external air pressure exerted on the outside of the container causes the container to cave in.

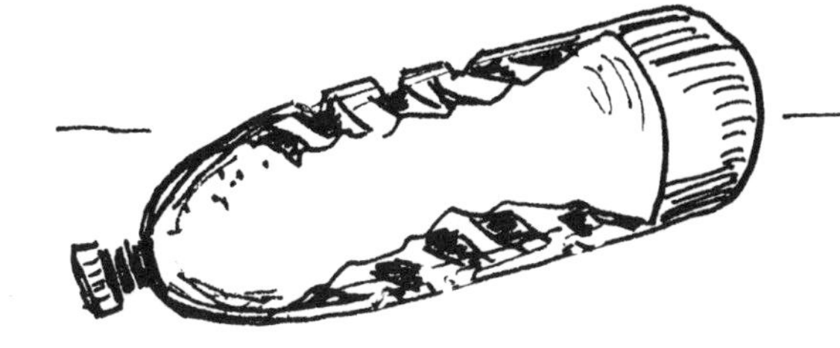

THE TWO-LITER RAIN GAUGE

To construct your two-liter, container rain gauge, cut the container into two parts. The bottom portion should be seven to seven and one-half inches high. The uncapped, funnel portion of the container should be inverted and nested into the bottom portion.

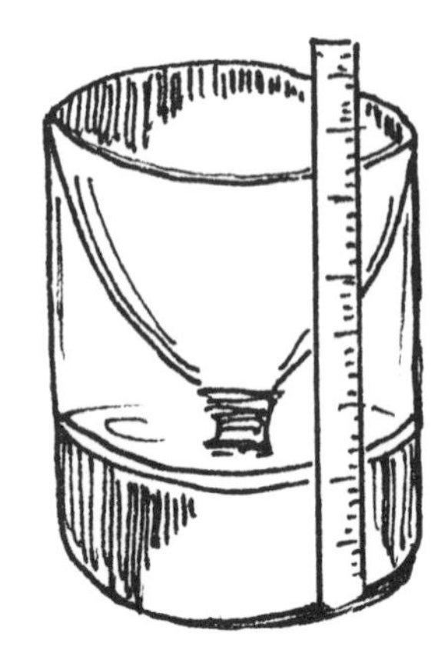

Use a ruler to mark a scale on the side of the container. Place your rain gauge in an open area removed from trees and bushes. If possible, keep it out of the wind. If the wind is a problem, a stake must be placed in the ground and the rain gauge fastened to it.

Record the rain that falls each day. Collected rain can be transferred to a graduated cylinder to determine a measure of the amount of rain water collected. Plot this on a graph. Remember to empty the bottom portion each day after you have recorded the amount of rainfall.

THE TWO-LITER, FREEZE-A-LITER

Interesting things can be investigated with two-liter, plastic containers wherein water is turned into ice.

Water is unusual in that it expands when freezing. At ordinary temperatures water expands when heated and contracts when cooled, just like other liquids. If you cooled water, it would contract until it reached a temperature of 39° ($4^{\circ}C$). However, if water is cooled more, it begins to expand and continues to expand until it freezes at $32^{\circ}F$ ($0^{\circ}C$). When it freezes it forms ice which is lighter than water and floats in water. Ice takes up about one ninth more space than it does existing as water.

Fill a two-liter container with one liter of water. Mark the water level with an indelible marker. With the cap off, place this upright in a freezer. When the water has turned to ice, mark the top level of the ice cylinder. How does this mark compare to the previous marked water level?

I WANNANO:

***** When the ice melts and returns to a liquid state, will the liquid level be the same?

***** What is the mass (weight) of one liter of water plus the container it is in? When water freezes solid, does the mass stay the same, increase, or decrease? Make an inference. Check it out.

***** When water freezes into a solid inside the container, under room conditions, how long do you think it would take for the ice to melt while retained in the container? Would a similar shape cylindrical block of ice exposed to room conditions on all of its external surfaces melt in the same amount of time, less time, or more time? Make an inference. How can you find out?

***** Fill two, two-liter containers each with one liter of water. With the caps off, freeze the contents of both containers. Expose both containers to room temperatures. Cap one container. Leave the remaining container uncapped. Will the ice in both containers melt in the same time? Try it.

THE TWO-LITER LUNG MODEL

To construct your two-liter lung model, you will need one, two-liter, plastic container, a plastic or glass "T" or "Y," one to two feet of plastic tubing, a piece of thin, rubber or latex sheeting approximately, seven inches square (this can be cut from a large balloon or purchased from pharmaceutical stores), and two, small balloons.

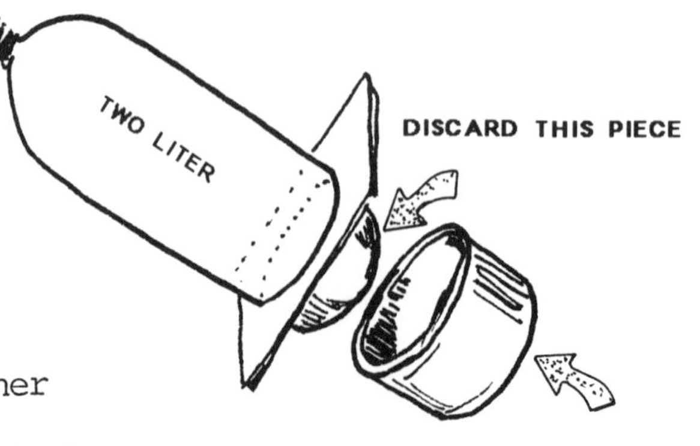

The two-liter lung model is constructed by pulling the black bottom portion off the container. Applying heat from a hair dryer to the bottom of the container while rotating the container will facilitate this spearation. Once the bottom section is removed, cut out the bottom portion of the container below the taper. Retention of a portion of the container's taper is necessary so that the container can later be reinserted into the black section of the container. From the black section, cut out a two and one-half inch diameter circle. This bottom piece fits back on the tapered end of the container. However, before slipping this bottom piece back, place a piece of thin rubber or latex sheeting (approximately seven inches square) over the tapered end of the container. This sheeting becomes locked in place when the black portion is reattached to the container.

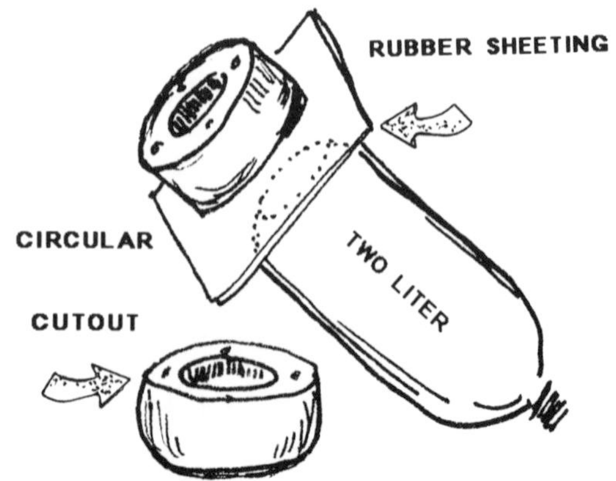

Prior to reattaching the bottom section, complete the inner portion of the container. Cut a hole in the cap (punch or drill) to accomodate the outside diameter of your plastic tubing. Most hardware stores stock a variety of plastic tubing. At both ends of the plastic "T" or "Y" attach small balloons. Balloons that have been previously filled once or twice with air work best. At the top of the "T" or "Y" fasten the tubing. If the joining of the tubing and the balloons is not snug, use a pipe cleaner or a twist tie and tighten all joints. Push all this through from the bottom, threading the plastic tubing through the container cap. Screw the container cap down tightly. If you have an air leak around the tubing as it comes in contact with the cap, seal it with soft clay or caulking compound.

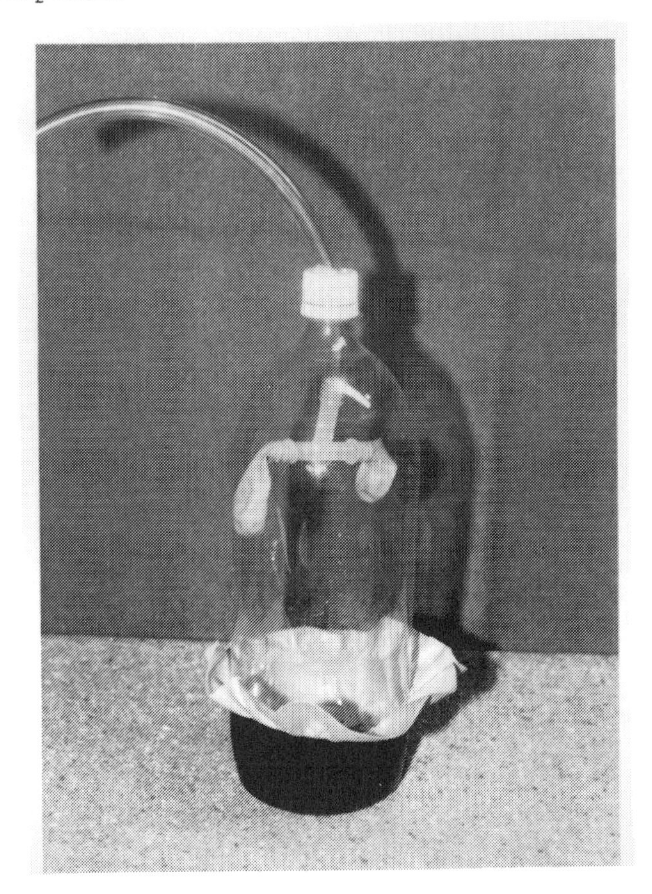

The "T" or "Y" represents the windpipe, the balloons the lungs, and the bottom container the thoracic cavity.

Air pressure inside the two-liter container (the chest) is reduced by pulling the rubber sheeting outward. Pinching off the plastic tube (windpipe) and pull and release the pressure on the rubber sheeting. What do you observe?

DOUBLE LUNG MODEL

A single, lung model can be constructed using the upper funnel-shaped portion of the two-liter container. Fold the lip of the balloon over the container's opening.

SINGLE LUNG MODEL

THE TWO-LITER, LUNG-CAPACITY TESTER

What is the capacity of your lungs? How much air do you exhale from one breath? How does the volume of this exhalation change with physical activity? This information can be determined by using the lung-capacity tester.

To construct the two-liter, lung-capacity tester, you will need two, two-liter plastic containers, several feet of flexible, plastic tubing, and an over-flow basin.

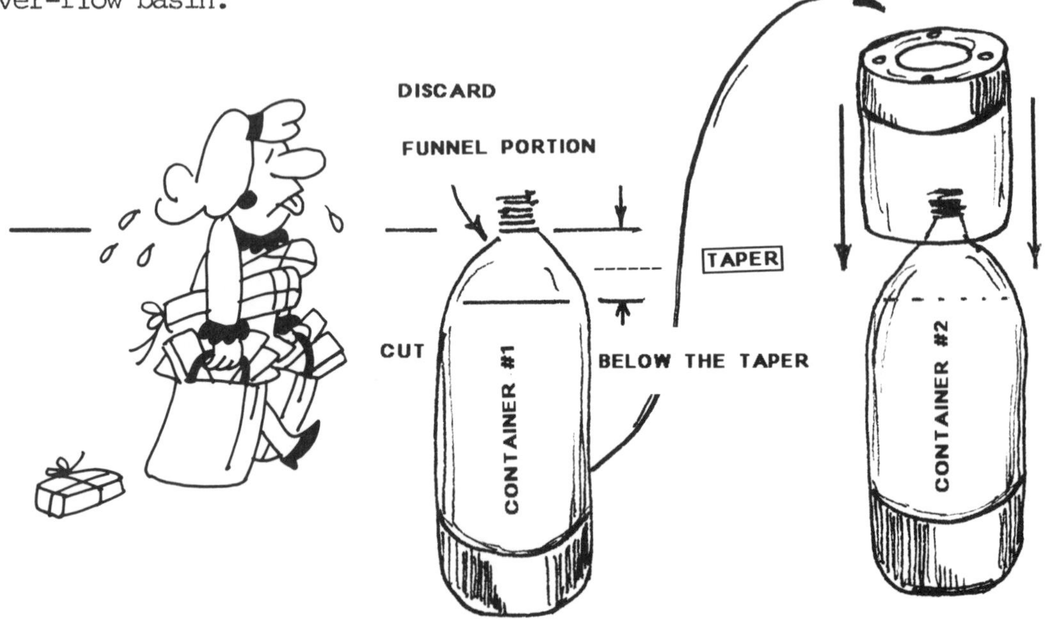

To make your lung capacity tester, cut the funnel top portion off one of the containers. The cut should include all of the upper, tapered portion of the container plus a small portion of the cylinder.

To the second container, add water in units of one-hundred, cubic centimeters. Make an indelible mark on the outside of the container for each graduation. This becomes your vertical ruler. Completely fill the container with water. With the water-filled container in an upright position, place the bottom portion of the other container so that it nests on top of the upright container. Push the bottom portion securely down over the filled container. Invert the entire apparatus. Some water will flow into the bottom portion. This situation will stabilize. Place the entire apparatus inside a large basin to catch any subsequent water runoff.

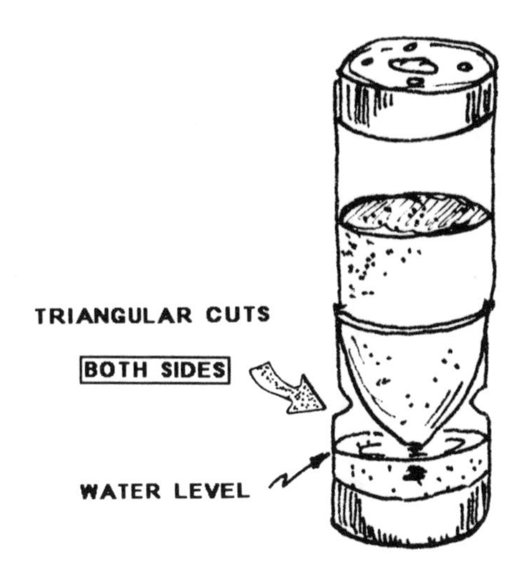

Make two or three, small triangular cuts in the wall of the bottom container. These cuts should be made well above the present water level. These are overflow outlets. When the triangular sections are cut out, additional water will leave the upper container. This situation will stabilize. Mark the level of the water on the outside of your container. Insert tubing through one of the triangular openings and thread it into the inverted container.

Exhale your breath through the external tubing leading into the water-filled container. This action displaces water inside the inverted container. When you have exhaled all of your breath, make a mark at the current water level. The difference between the two marks is the amount of water displaced by you and this represents your lung capacity. This can be translated into a volume measure by reading the one-hundred, cubic centimeter ruler you previously constructed. If an individual's breath exceeds the limits of one container, double the system connecting two, two-liter testers with an external plastic or glass "T."

I WANNANO:

***** Is lung capacity independent of gender differences?
***** Is lung capacity effected by age, weight, height, or athletic ability?

Note: For health reasons individual straws should be joined to the tubing's end.

THE TWO-LITER THERMOMETER

To construct your two-liter, container thermometer, you will need one, two-liter container, a soda straw, and a one-hole cork that will accomodate the insertion of the soda straw. Fill the two-liter container to the brim with water tinted by the addition of one or two drops of ink or food coloring. Cork the container securely. Insert the straw. If necessary use soft clay to seal the area around the straw. You should now have an air-tight, closed system.

Do not squeeze the water-filled, two-liter container thermometer. Compression on the cylinder wall will turn your thermometer into a water fountain. Using an eye dropper, add water to the straw filling it to a point where the water level in the straw is visible and will remain visible for small fluctuations in temperature. Allow the water in the container to reach ambient temperature. Mark the straw at the present water level. Equate this mark with the current known temperature. For example, if the room temperature is 70°F, the mark on your straw represents 70°F. This mark becomes your reference for increases and decreases in temperature.

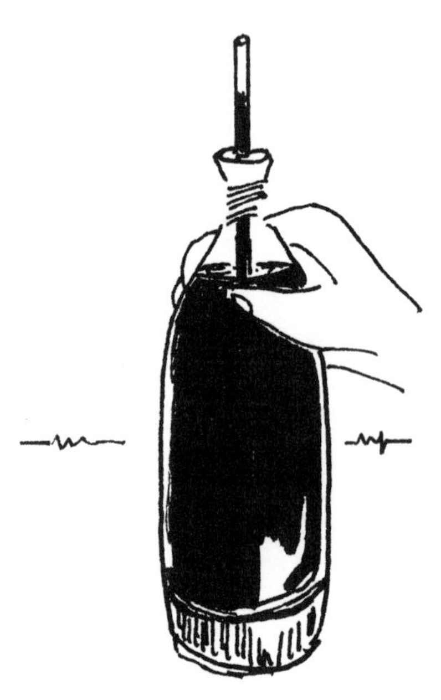

Water in the straw will rise when warmed. Water will drop in the straw when cooled. Additional marks can be made on the straw as the temperature either rises or falls. These marks can be recorded and equated with known temperatures. Once a few designated marks are made on the straw, you can infer temperatures from reading the notations marked on the straw.

For more immediate evidence of the expansion and contraction of water, immerse the two-liter thermometer in a pot of warm water. What do you observe?

THE TWO-LITER FLOTATION DEVICE

A capped, liquid-filled, two-liter, plastic container will sink when placed in water. A capped, empty, two-liter container will float in water.

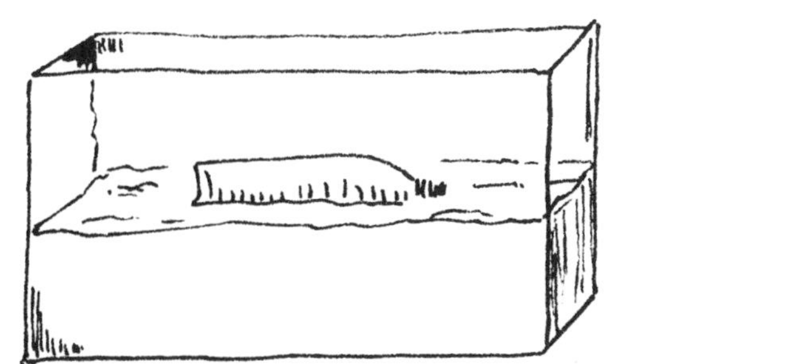

How much weight must be attached to the empty container to sink it?

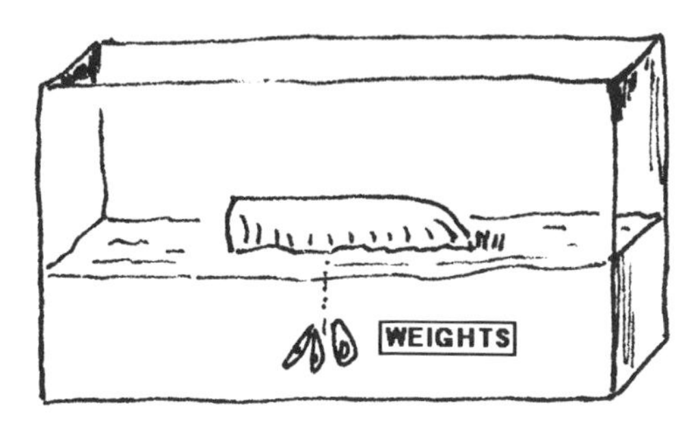

If two or more containers are joined together and floated as one unit, how much weight will it take to sink the containers?

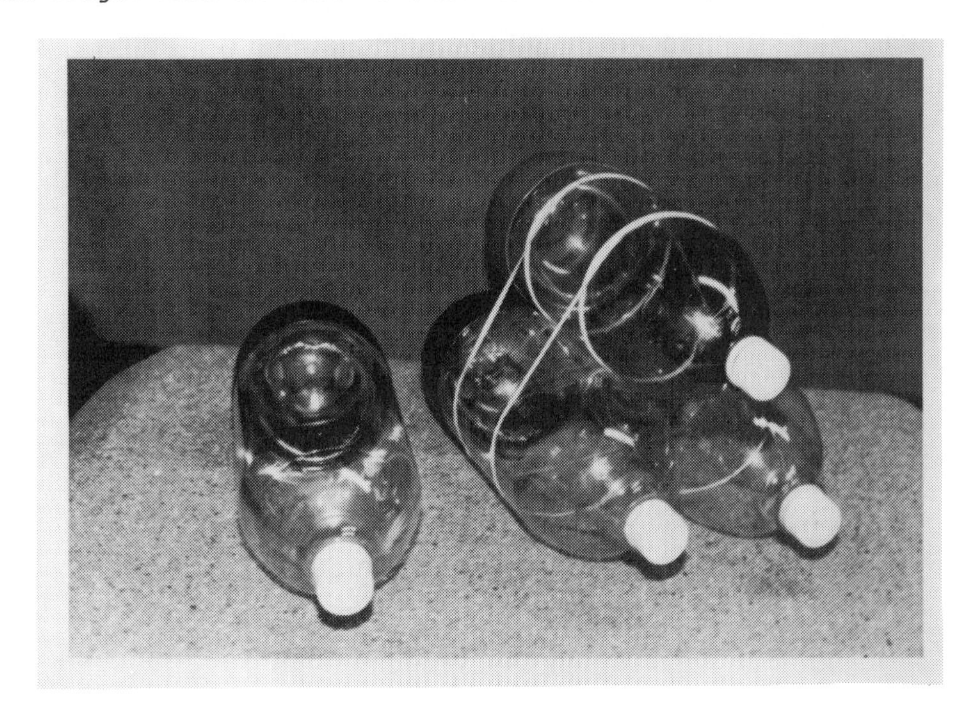

I WANNANO:

***** How many empty, capped containers would it take to support a one-hundred pound individual in water?

THE TWO-LITER GAS CHAMBER

Even a presumably empty, two-liter plastic container contains air. Air can react with items placed within the container. In some instances a gas may be generated. A closed container with a flexible balloon cap can reveal any expansion within the container. For example, place a small amount of water (25-35 cubic centimeters) in the bottom of a two-liter container. Drop an antacid tablet into the container. Quickly cap the container with a balloon. A leak proof, snug fit is necessary. Also, the contents of the container may need to be periodically, gently swirled. Observe what happens to the balloon.

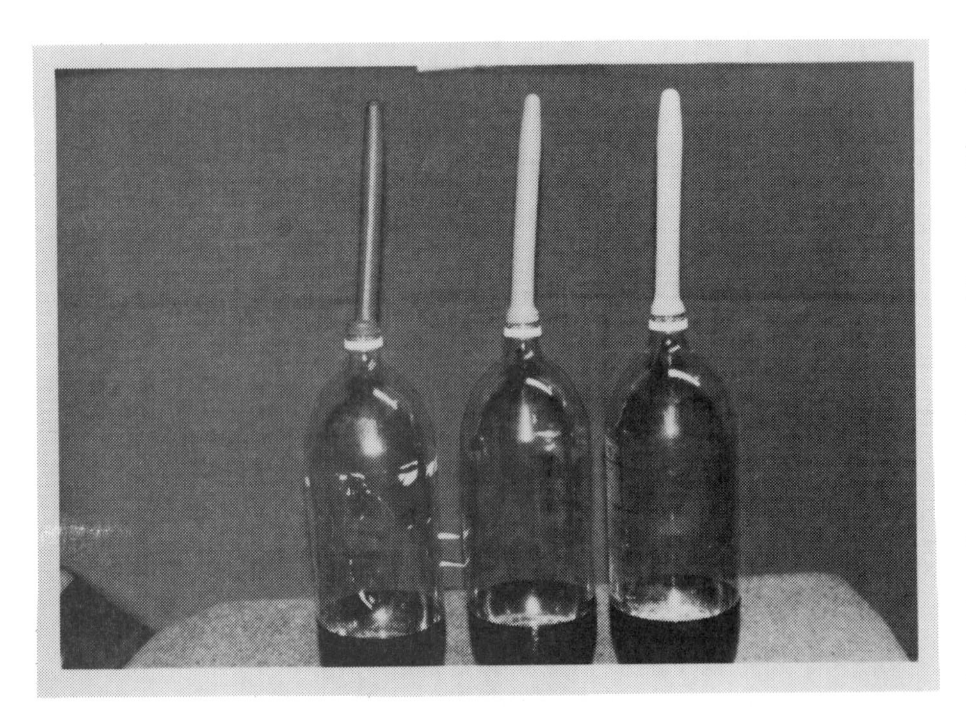

Expanding gases cause the balloon to be inflated. Repeat this process comparing portions of antacid tablets, for example, one-half of an antacid tablet, to a whole antacid tablet, to a one and one-half antacid tablet. Rank order the containers from the least balloon inflation to the most balloon inflation.

EXTENSIONS...

***** Do equal amounts of differing antacids inflate the balloons in the same manner?

***** Using three, two-liter containers, fill one, one-quarter filled, another one-half filled, and still another three-fourths filled with fresh cut grass clippings. Cap each container with a snug fitting balloon. Place all three containers in direct sunlight. Describe your observations.

THE TWO-LITER DENSITY APPARATUS

It is not only solids that sink or float in water. Liquids find their own level in relation to water based on their own unique densities. If a liquid does not mix with water, it is possible to determine if a liquid is more or less dense than water. The weight of a volume of a particular liquid may differ from the weight of an equal volume of another liquid. A base or point of reference is used to compare liquid densities. Water is often used for this purpose. As a base or point of reference, water has been assigned a density of $1.0 g/cm^3$, or simply 1. Liquids less dense than 1, float on water. Liquids more dense than 1, sink in water. And liquids that are equal in density to water barely submerge themselves in water and do not sink to the bottom.

Pour about two and one-half cups each of water, cooking oil, and molasses into a two-liter container. Shake the container. Put it aside and let the contents settle. These three liquids will separate themselves out based on their densities.

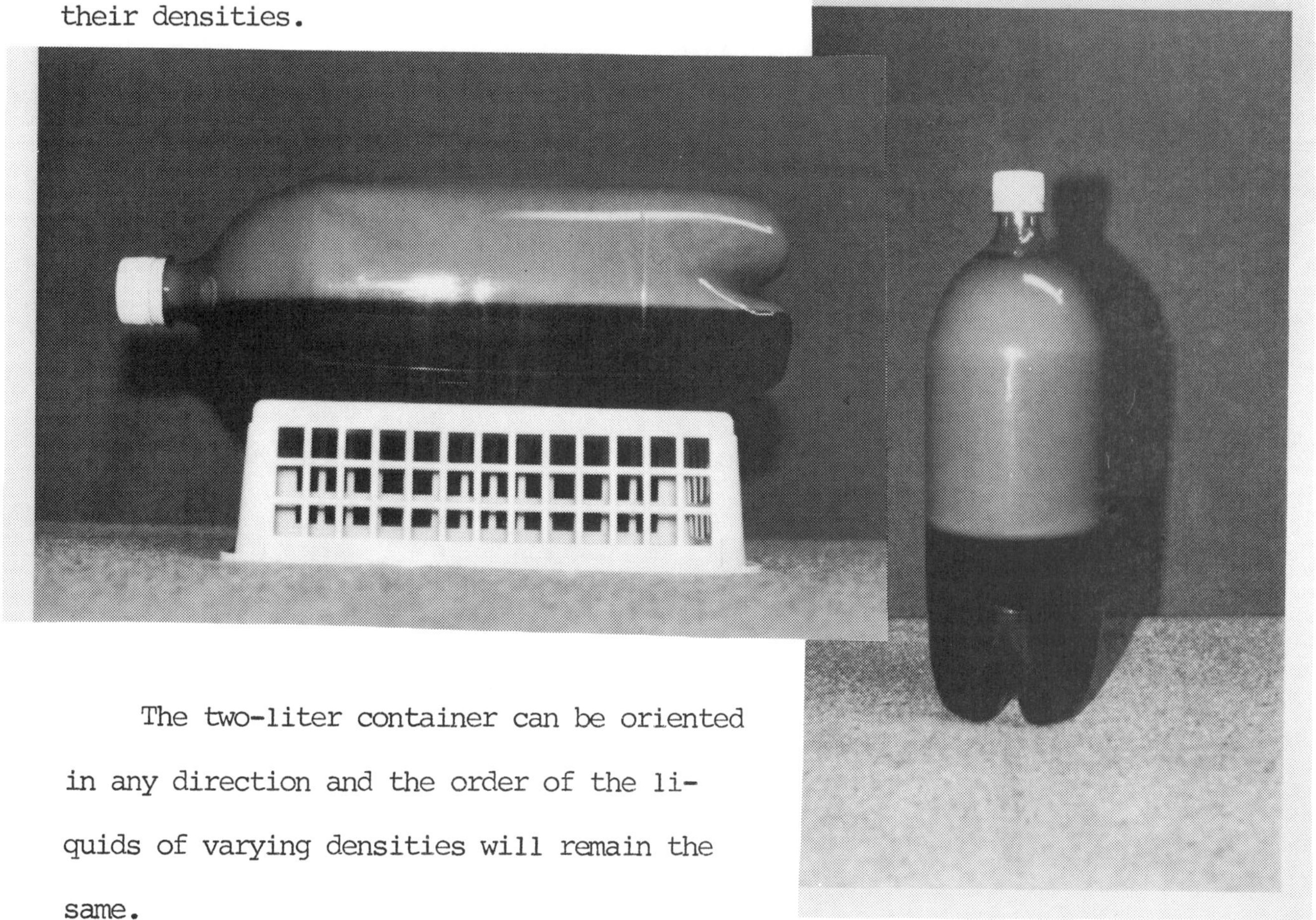

The two-liter container can be oriented in any direction and the order of the liquids of varying densities will remain the same.

***** Which of the three liquids (water, cooking oil or molasses) located itself at the bottom portion of the container?

***** Which liquid has a density greater than water?

***** Which liquid has a density less than water?

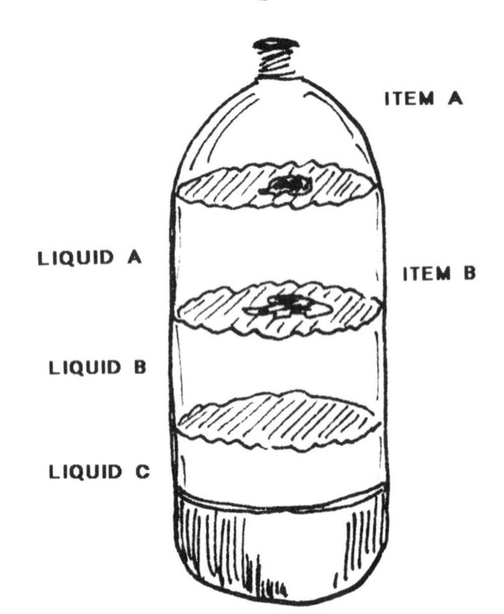

Some solid objects have a density less than the density of the liquid in which it is placed. These objects float on the surface. Objects with densities more than the density of the liquid in which they are placed sink to the bottom.

Float a piece of candle, a cork, and a paper clip in your three-tiered water, cooking oil, and molasses liquid layers.

***** Infer where the solids will be positioned within the system. Did your observations match your inferences?

***** Water can float on water. Mix a little food coloring into a glass of hot water. Within two inches from the top, fill your two-liter container with cold water. Using an eye dropper, slowly run some colored, hot water down the side of the container so that it spreads out slowly over the cold water. What do you observe? Continue to carefully add hot water until you have a layer about one inch deep. What happened as the hot, colored water cooled? Reverse the process. Add colored water to clear hot water. Describe your observations.

***** Place some cooking oil in a two-liter container. Use a sufficient amount of cooking oil to float a piece of ice cube. Describe your observations. How can you explain your observations?

THE TWO-LITER SPITBALL ACTIVITY

An empty, two-liter plastic container is not really empty. It is filled with air. Air exerts pressure. For all practical purposes the air pressure inside an uncapped, two-liter container is the same as the air pressure outside the container.

Roll up a small piece of paper into a pea-sized ball. Place this spitball in the neck of the container as close to the front of the container as possible. Attempt to blow the paper ball into the container.

The degree of difficultiness in accomplishing this task lies in the fact that as you direct a stream of air at the ball, you are indeed blowing into the two-liter container thereby increasing the air pressure inside the container. The pressure inside the container is now greater than the pressure outside of the container. In a sense, this is a self-defeating exercise inasmuch as the harder you blow the more the internal air pressure increases and the task becomes increasingly difficult.

THE TWO-LITER WATER FILTER

To construct your two-liter, plastic container water filter, cut the top off the container and invert this into the bottom portion.

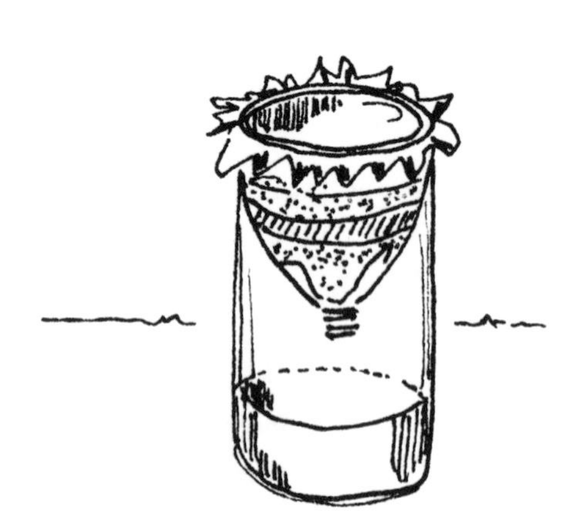

Insert coffee filter paper into the inverted cone. Add a layer of wet sand. On top of this add a layer of powdered charcoal and another layer of wet sand. Slowly pour in a portion of the muddy water sample. Observe and describe the liquid that collects in the bottom portion of your container.

WARNING: Do not drink the collected, filtered water as it may contain harmful bacteria.

***** How could you improve on the above process of water filtering?

THE TWO-LITER SOLAR TRAP

An empty, two-liter, plastic container can be an effective solar trap. Remove the cap from the container. Gently lower a thermometer to the bottom of the container. Record the temperature inside the container. Compare this reading to the temperature outside the container. Place the empty container anywhere you like and in any position you like to increase the temperature inside the container. Try this again comparing two containers, positioned in the same manner, comparing one capped container to one uncapped container. What do you observe?

THE TWO-LITER WATER POWER APPARATUS

You can construct your two-liter, water-power apparatus by punching three holes, all of the same size, in a verticle line one above the other. These holes should be positioned at approximately one quarter, one half, and three fourths of the container's height.

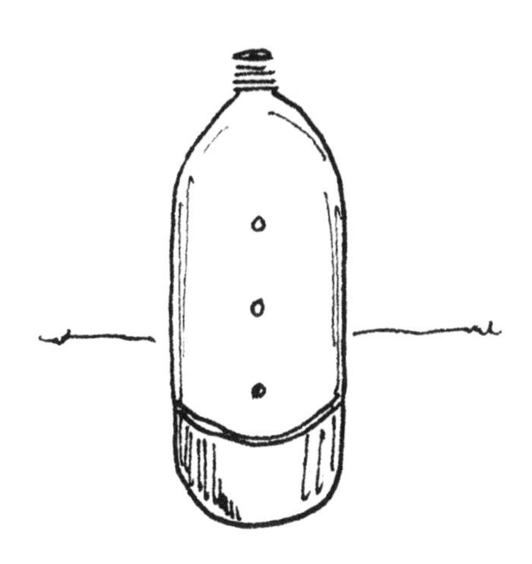

Place a small square of scotch tape over each hole. Fill the container with water. Put the cap back on the container. Over a wash basin, remove the tape covering from all of the holes. Gently squeeze the container. Observe the reaction. Squeeze harder. Again observe the reaction.

The water at the bottom of the container is pushed harder by the weight of the water above it. It produces a stream longer than all the rest. Refill the container. Invert the container. Squeeze it. Do the results differ?

THE TWO-LITER PENDULUM

Two-liter, plastic containers lend themselves well to investigating pendulums. Three major conditions, relative to what influences the motion of pendulums, are usually investigated. These are the mass of the pendulum bob, the length of the string that supports the pendulum bob, and the angle from which the pendulum bob is released. In experimenting with pendulums, little difficulty exist with altering the length of the support string or the angle of release. However, it is difficult to obtain objects, for example, spheres, that have the exact shape and volume but which possess different masses. The shape and the volume of two-liter containers are all the same. However, their interiors can readily be altered by adding sand or water thereby providing bobs of similar shapes and volumes but dissimilar masses.

The period of a pendulum depends upon the length of the supporting string and is independent of the mass (or weight) of the pendulum bob, and to a limited extent, is also independent of the amplitude angle of release. All this can easily be verified. Try it!

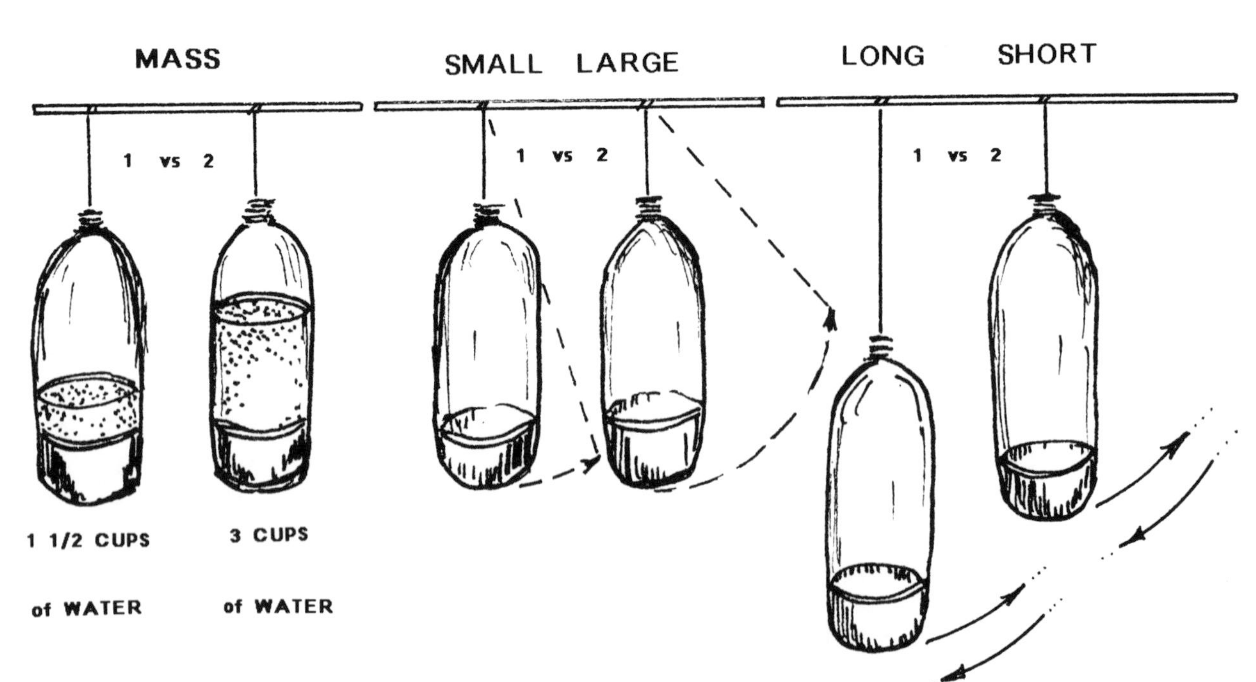

***** Cut, two, equal lengths of string (approximately one meter each). Punch a small hole in each of two, two-liter container caps. Insert the string through the hole and tie it around a small portion of a toothpick or match stick so that it cannot be pulled out under pressure. Fill one, two-liter container with one and one-half cups of sand or water. Hang the pendulum bob pendant from an overhead crossbar. Gently set the pendulum in motion. Time how long it takes to swing back and forth ten times. Repeat this same process using the pendulum bob containing three cups of sand or water. Infer how the recorded times will compare. Observe and record your results. What did the differences in the mass of the bobs have on the motion of the bobs?

TRANSFER OF ENERGY PENDULUM

Energy transfer can be shown by using coupled pendulums. Fill two, two-liter containers with equal amounts of sand or water. Cut two pieces of string about one and one-half to two feet long. Thread the strings through pre-drilled holes in the container caps. Affix the strings to a portion of a toothpick or match stick. Screw the caps on the containers. Hang the two, two-liter containers pendant from an overhead crossbar (broom handle) making sure the containers are equal distance down from the crossbar. Notch both ends of a soda straw so that the strings can be wedged into them. Move the two, two-liter containers close enough so that the soda straws can fit between them.

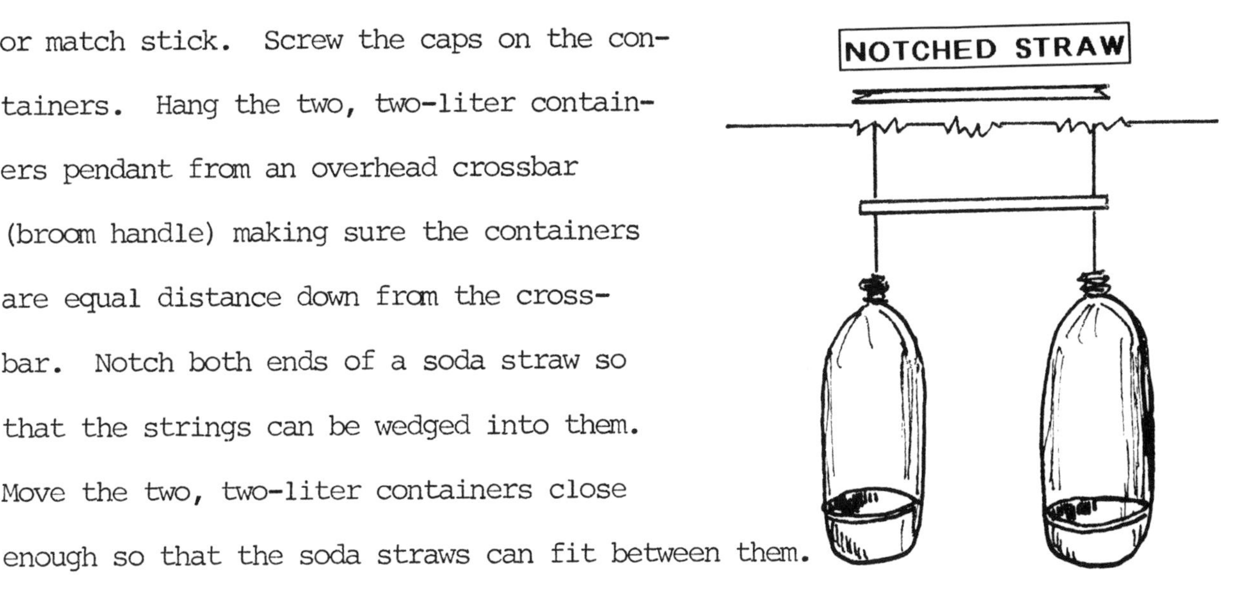

Release the first pendulum bob. Make an inference as to what will happen. Observe. Offer an explanation for your observations.

***** Repeat this activity placing dissimilar masses in the two-liter containers. Again, make an inference as to the anticipated results. What do you observe? What do you conclude?

THE TWO-LITER SAND PENDULUM

In 1851, the French physicist J. B. L. Foucault established by a simple experiment that the earth really rotated about its axis. This he accomplished by hanging a long (200 feet), heavy (62 pound cannon ball) pendulum from the dome of the Pantheon, a public building in Paris. Foucault's pendulum seemed to swing in an arc, indicating that the earth rotated under it. These conditions are difficult to duplicate. Nonetheless, some notion of the earth's motion can be garnered from observing the sand pendulum. Interesting sand patterns can be observed from viewing the results of the swinging sand pendulum.

To construct your sand pendulum, you will need one, wire clothes hanger, one, two-liter container with a small hole drilled or punched through its cap, and a quantity of fine sand. Commercial sand-box sand is too coarse and does not work well. Pet supply stores carry a variety of fine sand in a host of different colors. You must match the grain size of the sand to the aperture you made in the container cap. The sand pendulum is held pendant by the wire, clothes hanger.

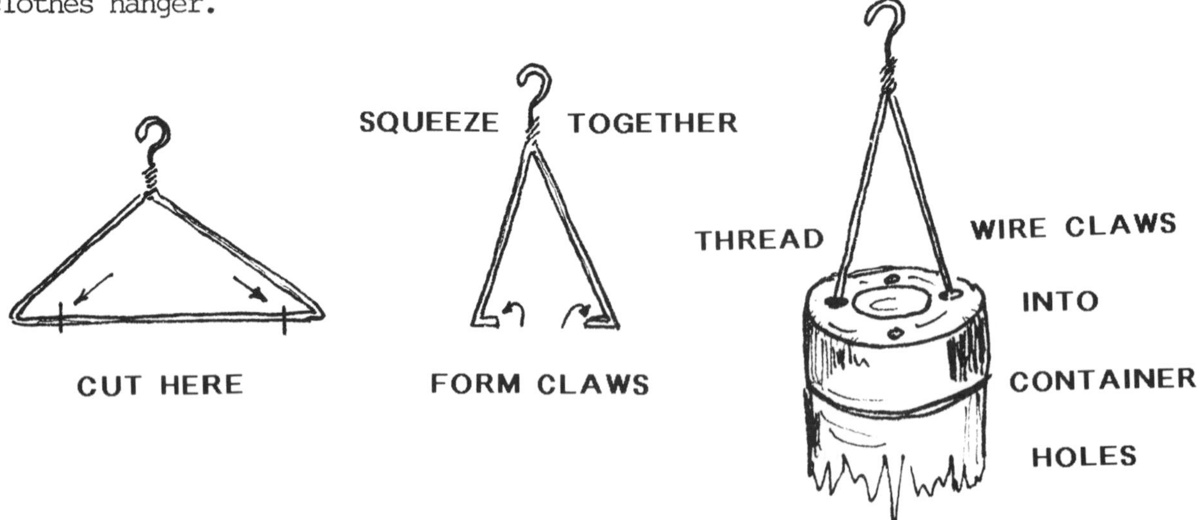

Fill the container approximately one-third full with fine-grained, colored sand. Replace the container cap. Cover the hole in the cap with a small square of scotch tape to temporarily seal it. Invert the container. Attach the wire clothes hanger. Suspend this apparatus from a string. Spread newspapers to collect the released sand as it forms different patterns. The newspaper allows for the easy pickup of the sand for cleanup or recycling purposes.

Pull the pendulum out to the desired angle. Remove the scotch tape. Observe the resultant pattern.

Vary the release swing, for example, straight plane or circular motion. What factors influence the generation of specific sand patterns?

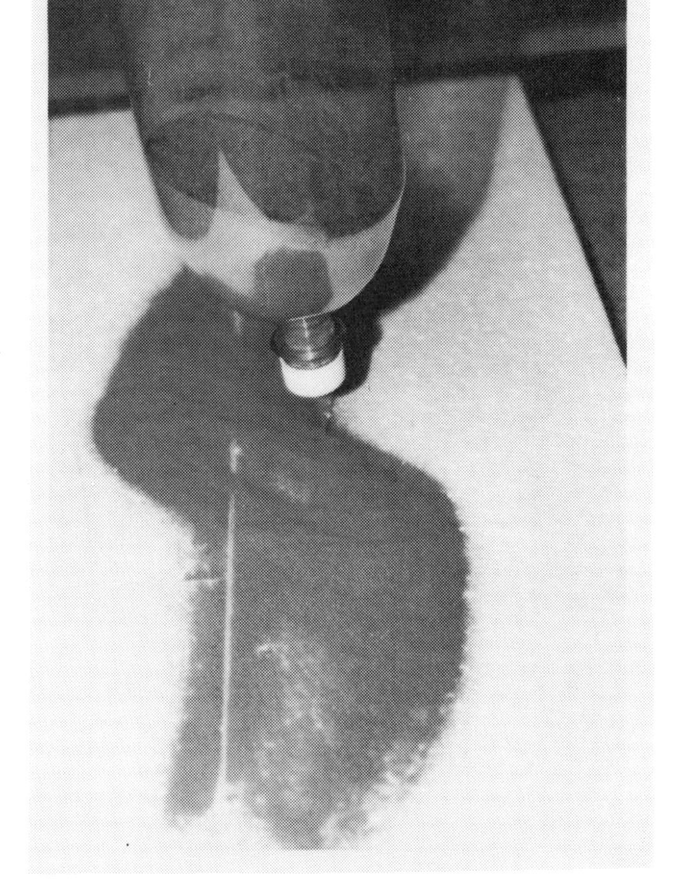

THE TWO-LITER POROSITY MEASURING APPARATUS

Not all solids are as solid as they appear to be. For example, granite and sandstone rocks are solids. Yet, they are different in numerous ways. One way has to do with the structure of the grains that make up these rocks. The individual grains that make up granite are interlocking. These grains are difficult to separate one from another. The individual grains that make up sandstone are non-interlocking. These grains are readily separated one from another. Rocks with interlocking grains do not allow liquids to pass through them. Rocks with non-interlocking grains have interconnecting spaces or pores between individual grains. Liquids can pass through these interconnected openings.

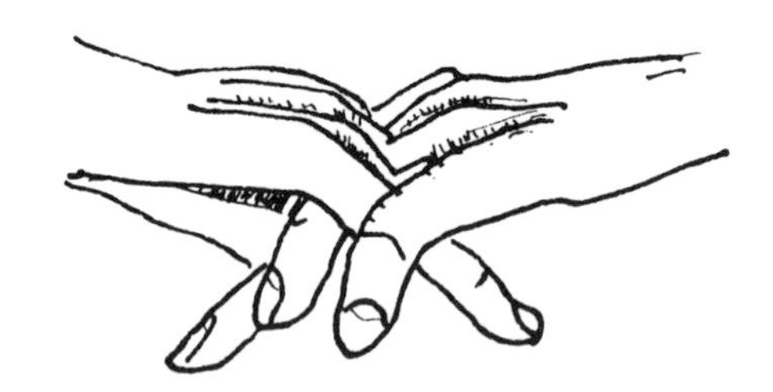

Interlocking Grains

Grains are locked to one another by the way they are linked together. Fold your hands. Interlock the fingers. Have someone try to pull your hands apart. Is it difficult or easy for them to separate your hands?

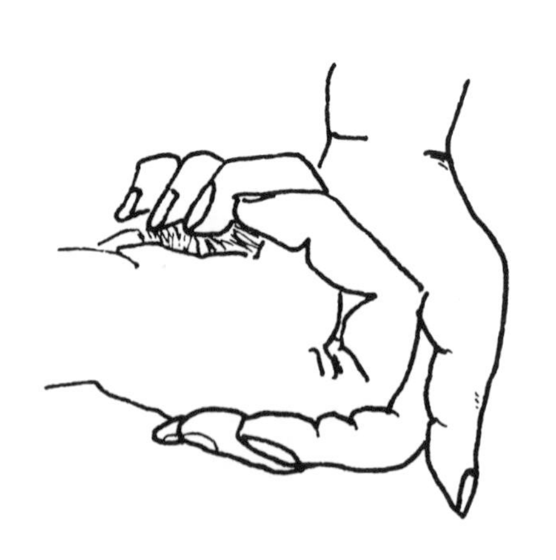

Non-interlocking or Separate Grains

In contrast with interlocking grains, we have non-interlocking or separate grains. These are isolated grains that are loosely cemented together. Fold your hands. Place a closed fist inside your other open hand. Close this hand about the fist. Have someone try to pull your hands apart. Is it easier or harder now to separate your hands then when your fingers were interlocked?

The ability for a liquid to pass through a material via these pore spaces is called porosity. The pore-space volume of a material can be measured.

***** Completely fill one, two-liter container with pea-size gravel. The volume of the two-liter container is known. Using a container to which you have added a known quantity of water, pour water from this into the two-liter, gravel-filled container until the water fills all the openings or pores between the gravel particles. Subtract the measured amount of remaining water from the measure of the initial amount of water. This difference is equal to the volume of the total, pore space between the gravel particles. Remember one milliliter is equivalent to one cubic centimeter.

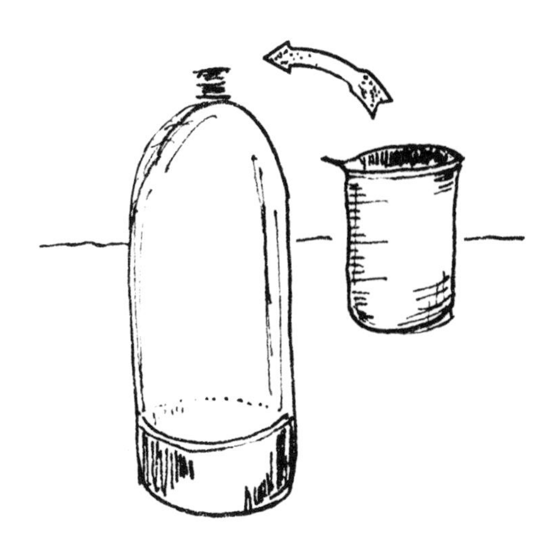

***** Fill another two-liter container with sand particles up to the same level as you filled the previous container with pea gravel. Slowly add water until no more water will soak into the sand and water begins to collect on the surface of the sand. Make sure all the sand in the container is completely saturated. Calculate the amount of water needed to saturate the sand thereby filling all the pore spaces between the sand particles. Subtract the numerical value of the amount of water remaining in the container from the initial numerical value. This amount is the total volume of pore space between the sand particles. How does this amount of pore space compare to the total, gravel pore space? What accounts for any differences?

THE TWO-LITER SERIES AND PARALLEL ELECTRICAL CIRCUITS

The process of manipulating batteries and bulbs seems magical to students. They are intrigued with hands-on, electrical activities because of their simplicity and sophistication. A study of elementary circuitry seems to engender challenges for students at all levels.

"How to set up series and parallel circuits using two-liter containers" is a high-interest, simple activity. However, it does take some basic skills with soldering and some careful prior preparation. As with all hands-on activities, challenges are more readily accepted if the initial engagement provides positive results. Therefore, all batteries, bulbs, and electrical connections need to be checked as to their usefulness prior to their inclusion in activities. Dead batteries and bulbs can tarnish lessons. Also, batteries and bulbs need to be checked as to their compatibility one with the other. Incorrect mating of bulbs to batteries blows them out or gives sub-standard performances. Most hardware stores can furnish charts that match battery size and strength to bulb type, size, and strength.

Batteries are the energy source. Bulbs house the wire filament that resist the current that passes through them from the energy source. The resisting, internal, bulb wire heats up and glows providing light. Household bell wire or something similar is needed to connect batteries to bulbs. Wire connections need to be as fool proof as possible to insure a problem-free lesson. Loose wire connections can be a source of frustration.

The use of some inexpensive equipment is necessary to insure success in the teaching of circuitry. You will need an inexpensive soldering iron, soldering paste, bell wire or similar electrical wire, a pair of wire cutters and strippers, a half dozen or so small, alligator clips, and the appropriate number of bulbs and batteries. The soldering iron should cost less than ten dollars. The cost of the remaining items, devoid of the bulbs and batteries, should cost approximately seven dollars.

WIRE CUTTER

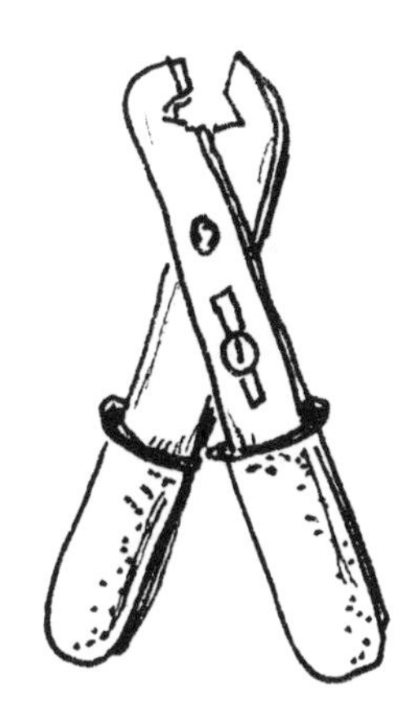

AND STRIPPER

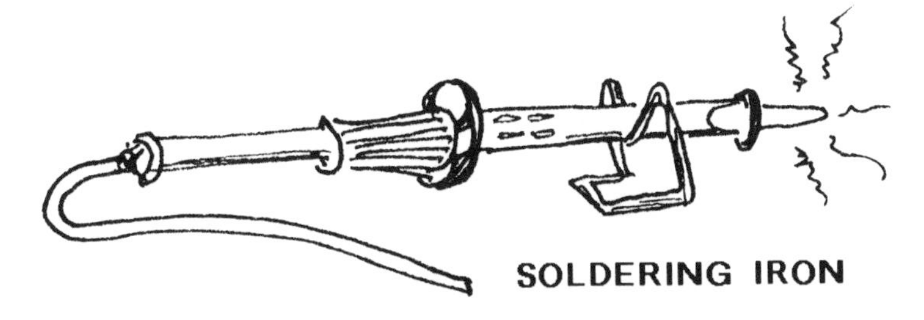

SOLDERING

Soldering is the act of permanently joining metal together. The solder, termed "soft" solder, is a mixture of lead and tin in varying proportions. Soldering works best when the objects being joined are clean, free from dirt or oxidation. A small piece of steel wool rubbed on the parts being joined will make the surface clean and bright and ready to accept solder. Soldering is a heat process. The soldering iron must be applied directly to the parts being joined. The soldering iron transfers its heat to the material being joined. The soldering wire is then applied to the joint not to the soldering iron. The heat of the wire being joined will melt the solder and it will flow to the designated area. Prior to applying the soldering iron to the area to be soldered, one should brush the area with a light coat of flux to assist the melting solder to flow smoothly. Flux also removes tarnish and oxides, and prevents them from forming. When joining two, insulated wires together, insulation must be removed prior to soldering. Wire strippers not only cut wire, but they strip wire of its insulation. You can cut the wire by inserting it into the rear of the jaws and squeezing. To strip the wire, you must use the notched portion of the wire stripper's jaw.

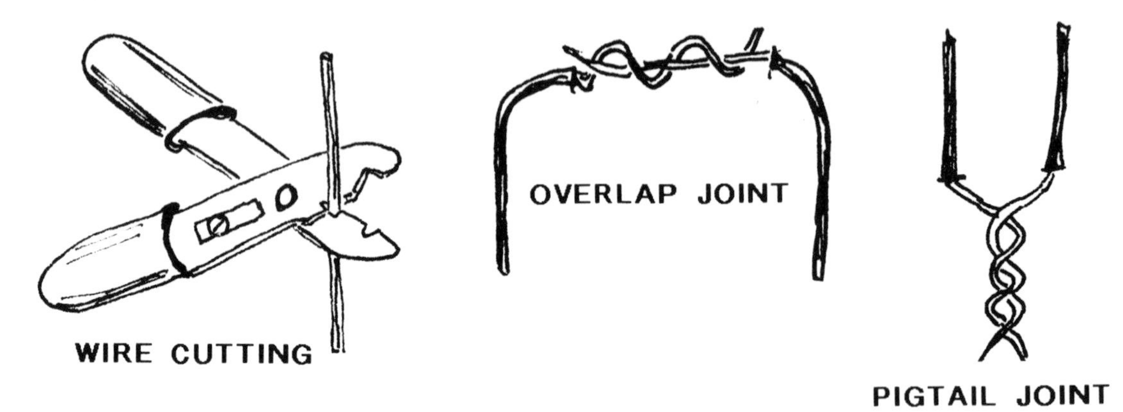

Soldering two wire ends together is enhanced by intertwining the wires prior to soldering. This can be accomplished in a simple overlap joint or a pigtail joint. A tight, wire wrap is essential to a good soldered joint.

When soldering, hold the point of the soldering iron on the area to be soldered. Do not press down hard. Heat accomplishes the task, not pressure. Let the iron work for you. The heat from the iron is transferred to the soldering area. This heat will cause the flux to melt, bubble, and flow. When the solder melts, slide the solder and the soldering iron along the area being joined. The iron, with its heat will move the solder much like liquid paint. When it is in this condition, cover all contact areas with a light coating of solder. A little solder, evenly distributed, is preferred to isolated globs. When soldering is complete, do not move the joined portions too readily. Solder requires a short period of time to cool down and solidify.

Occasionally, a soldering iron must be cleaned of accumulated solder and flux. Clean the tip with steel wool or sandpaper and coat it with rosin flux.

TWO-LITER CIRCUITRY

Two-liter container, bottom sections are used as bulb holders. The removal of these from two-liter containers is discussed earlier in the book. In an inverted order four holes are visible. Any or all of these holes can be used as bulb holders or sockets. These holes may need to be enlarged to snugly accomodate the bulbs you use.

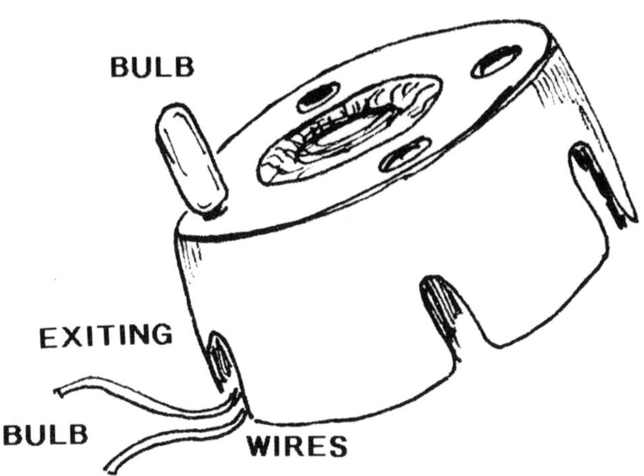

At each bulb holder, on the vertical cylindrical wall of the container bottom section, cut slots to allow openings for for exiting bulb wires. Each bulb must have two wires soldered to it. Snugly wrap one, stripped, wire end around the shank of the bulb. Solder this wire to the bulb shank. Solder another stripped, wire end to the base of the bulb. A small circle formed, using a needle-nosed set of pliers, in the end of the wire makes for easier soldering. For clarity use different colored wire for the shank joint and the end joint. As you assemble two or more bulb wire, hook ups, consistently use the same colored wire. When soldering, use the minimum of heat necessary to join the wires. Prolonged dwelling of the soldering iron in the joining of wires to the bulb, can render the bulb useless. Practice will insure success.

BULB-WIRE HOOKUP

RED-COLORED WIRE

WHITE-COLORED WIRE

SOLDERING COIL

Make as many of these bulb-wire hookups as you anticipate using. For one group of students utilizing one series and one parallel hookup, you will need at least eight, such bulb-wire attachments. If you make additional bulb-wire attachments available, you will invite open-ended discoveries on the part of the students.

SERIES HOOKUP

In a series hookup, the energy from the battery passes through the bell wire and through the wire in each individual bulb. This forms a continuous, uninterrupted flow. Any interruption, for example, poor connection or a dead bulb, causes the system to shutdown. Once the circuit is broken all the bulbs will go out.

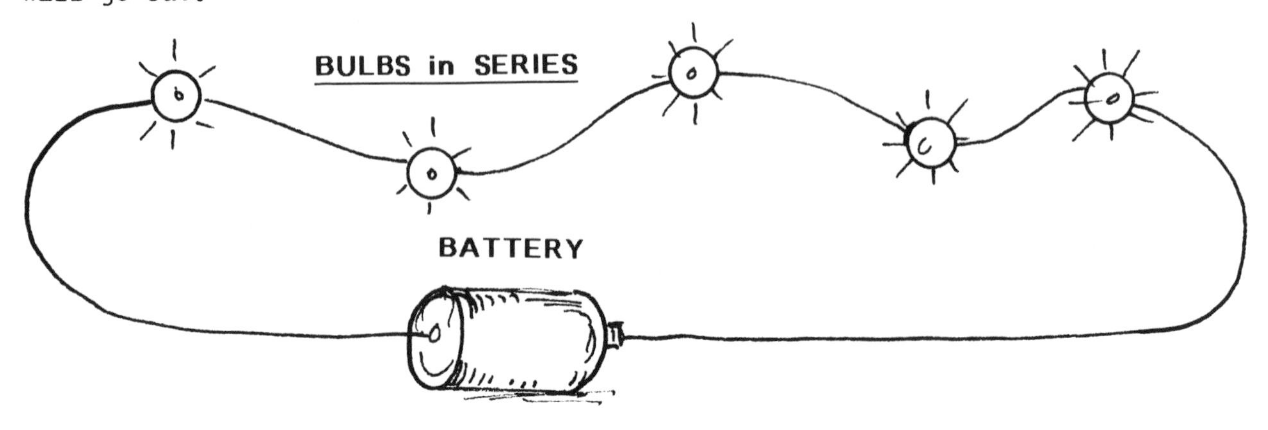

PARALLEL HOOKUP

By contrast, a parallel hookup is just that, a set of parallel wires which in and of themselves, form a circuit. In a closed system, devoid of bulbs, the current completes a circuit and electricity passes through the wire. Bulbs inserted into such a parallel system bridge the parallel wires enabling the electrical current to go two ways -- around the bulb in its original circuit and across and through the bulbs. Hence, when one or more bulbs in a system goes out, any remaining bulbs continue to shine brightly for the circuit is not interrupted or broken.

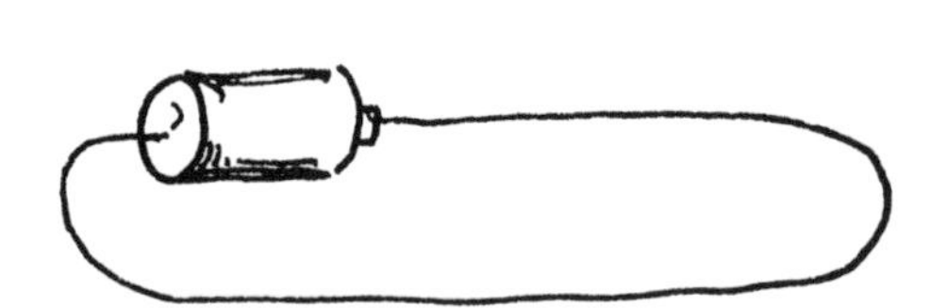

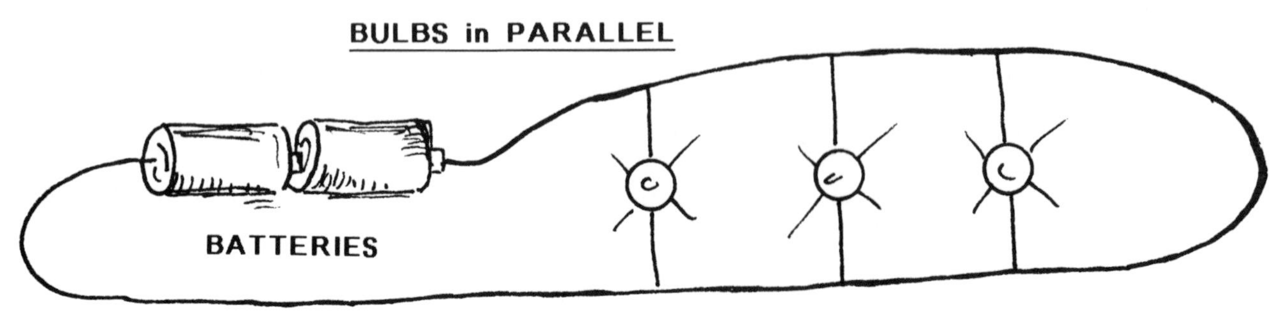

TWO-LITER SERIES HOOKUP

Insert a bulb-wire connector through each of the four holes. Extend the wire through the individual bulb slots.

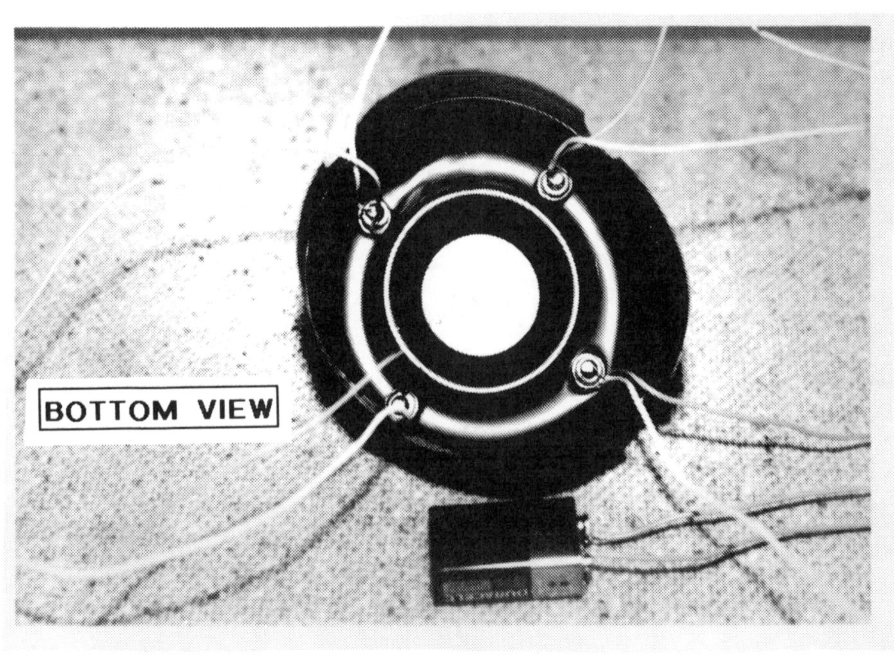

Pigtail join each bulb's red wire to the next bulb's white wire. Continue this for all but one set of red and white bulb wires. This open set will be connected to the energy source or battery.

Many times it is helpful to attach alligator clips to wire ends. They make it easy to join two or more wires. If you feel comfortable with your soldering skills, simply solder independent wires to the plus and minus terminals of your battery and attach alligator clips to the ends of these wires. This provides an easy, positive method of clamping two or more wires together.

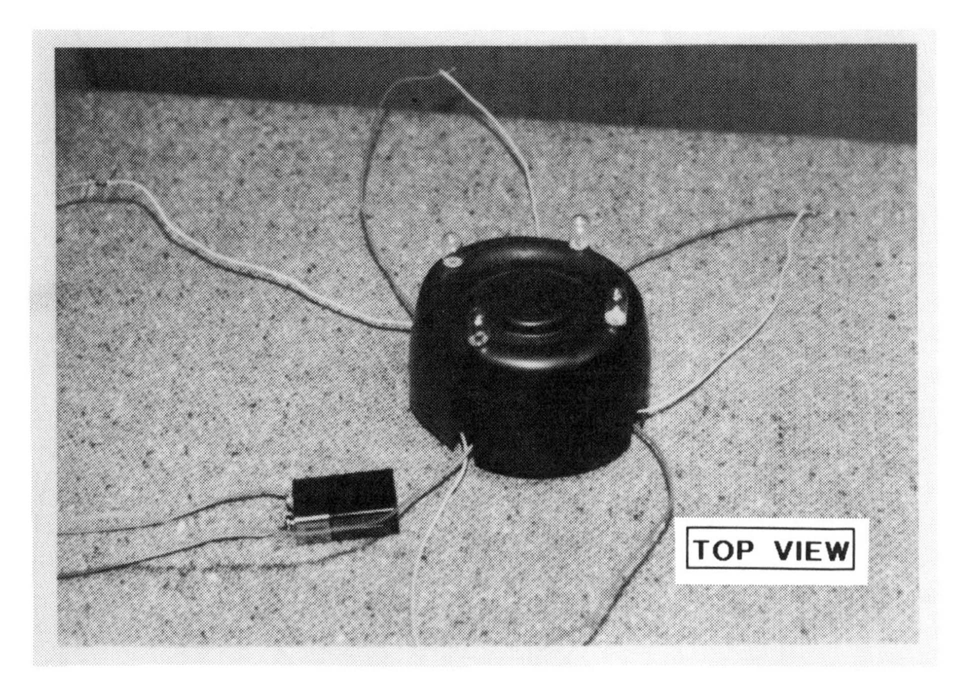

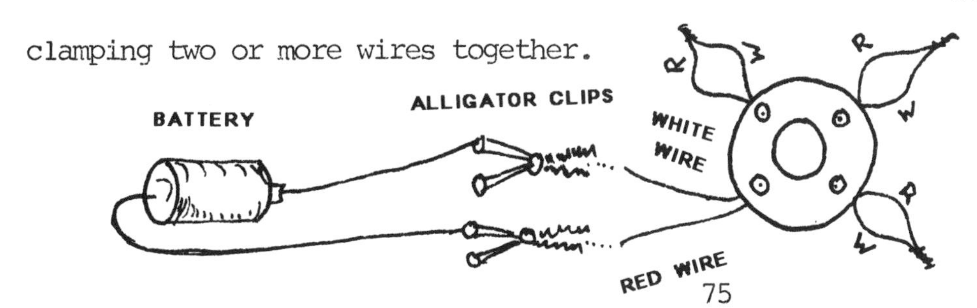

A TWO-LITER SERIES HOOKUP

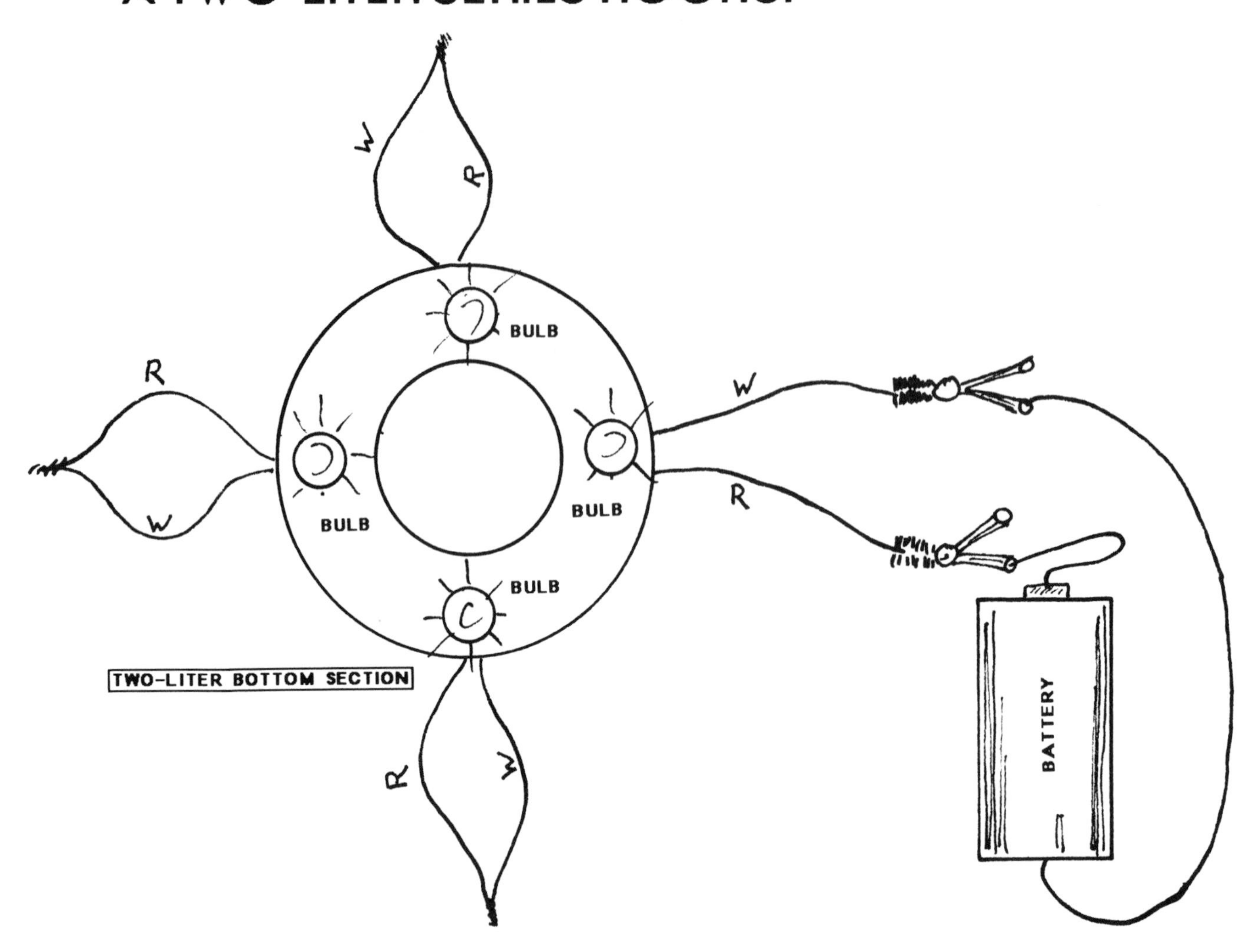

A TWO-LITER PARALLEL HOOKUP

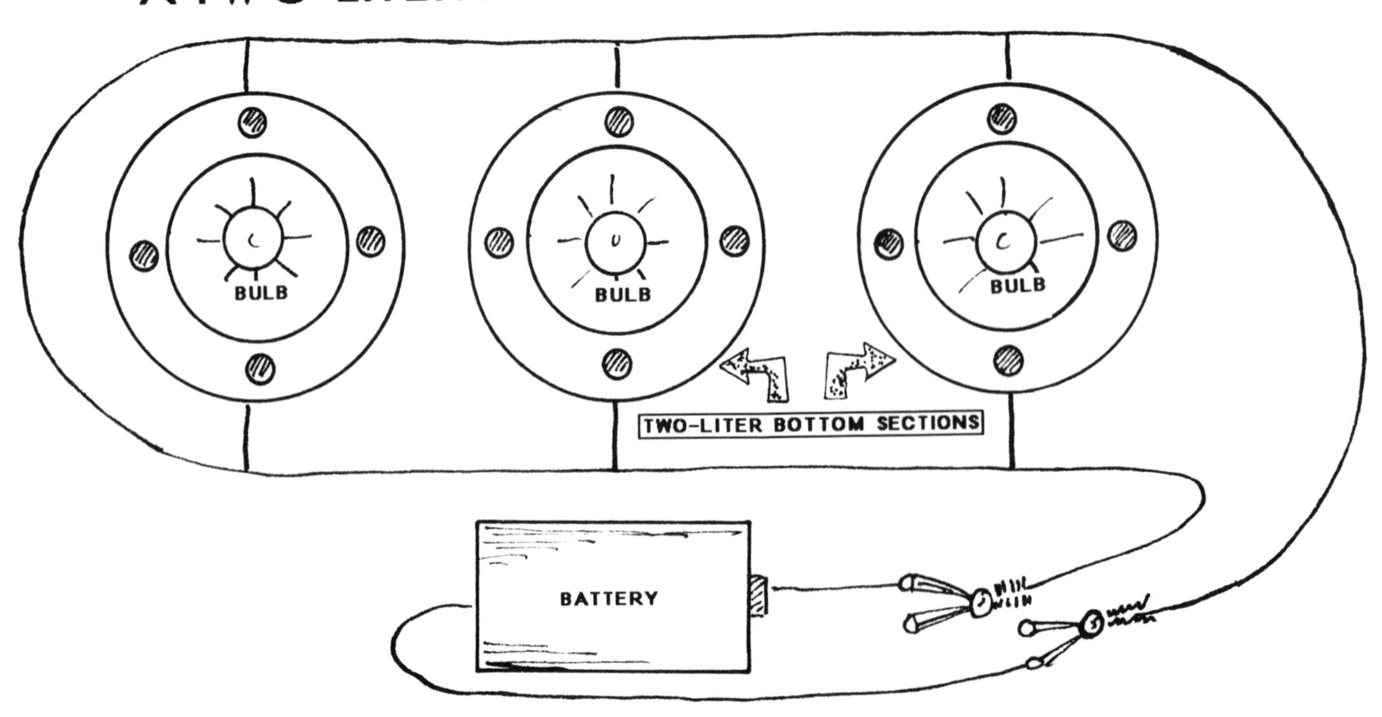

***** Compare the brightness of the bulbs in both circuits. What similarities and differences do you observe?

Should the lights in your home be connected in parallel or series?

***** If bulbs can be arranged in series and parallel, can batteries be hooked up in series and parallel? What would be the effect on the bulbs?

THE TWO-LITER FLASHLIGHT

To construct a two-liter container flashlight, you will need two, two-liter containers, two "D" size batteries, one bulb, some bell wire, two brass brads, one paper clip, three, terminal wire connectors, a piece of aluminum foil, and some soldering materials.

Taking the necessary precautions when working with sharp cutting instruments, cut into one, two-liter container making two, half-moon cutouts.

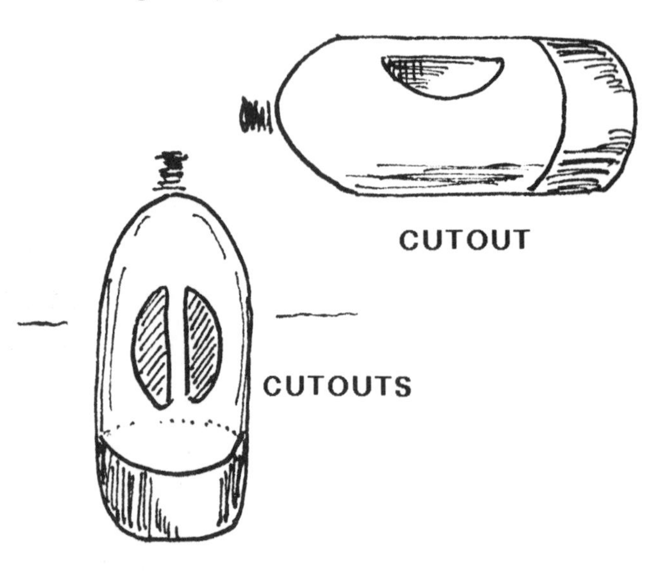

These cutouts are necessary when batteries and wire connections need to be made thereby completing the circuit. When the two-liter container flashlight is completely assembled, these openings also serve as a handle.

Punch or drill a hole in a container cap. This hole houses the bulb. Prior to inserting the bulb into the hold, solder two, six-inch lengths of bell wire to it. Strip away three-fourths to an inch of insulation away from both ends of the wire. Solder one wire to the shank of the bulb, and one wire to the bottom of the bulb. When this is

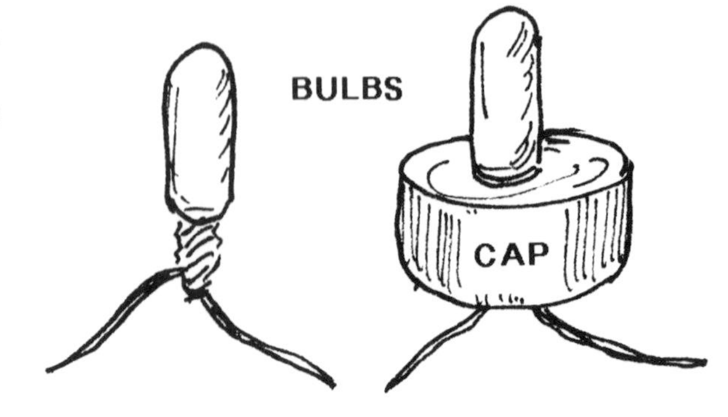

completed, insert the wire through the pre-drilled hole in the container cap. The bulb should nest snugly in the hole.

Cut the top off the second container. Remove the container cap. Using one of the "super" glues, glue the reflector funnel onto the cap. Line the funnel with aluminum foil to serve as a reflector.

To the top of one battery, solder a four-inch length of bell wire. To the bottom of the other battery solder a ten-inch length of bell wire. Join the two batteries together by taping them tightly together at area "A." Sustained contact between these two batteries must be maintained.

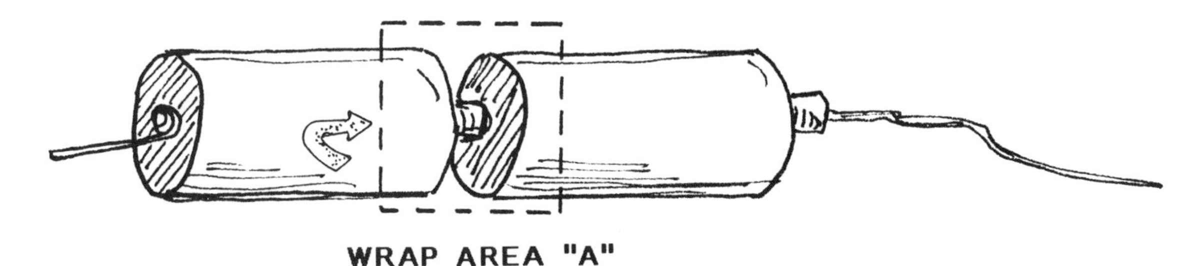

WRAP AREA "A"

Solder a six-inch length of bell wire to one arm of each brass brad. Punch two, small holes, one inch apart in the wall of the two-liter container.

Insert the brass brads and their corresponding soldered wires into the holes. Thread a small, paper clip under

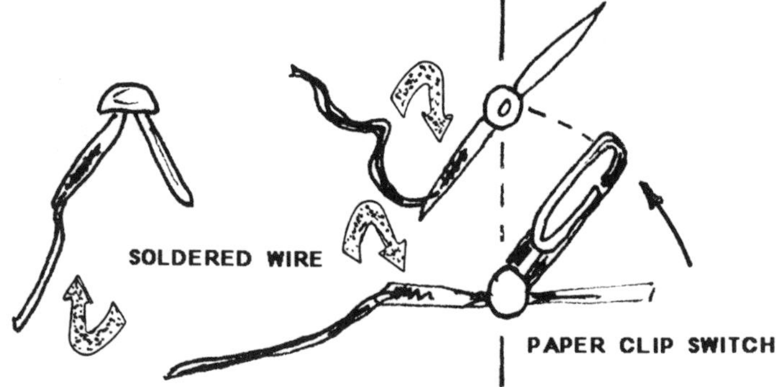

the head of one of the brads prior to inserting it into the hole. This paper clip should be free to swing back and forth.

Upon making contact with the other brass brad, the paper clip serves as a switch turning the flashlight bulb on or off.

Place the two, "D" batteries with attached wires under the handle of the two-liter container. Wrap tape around both the batteries and the handle of the container. This should be a firm, tight wrap. Three sets of unconnected wires need to be connected. Connect these as shown.

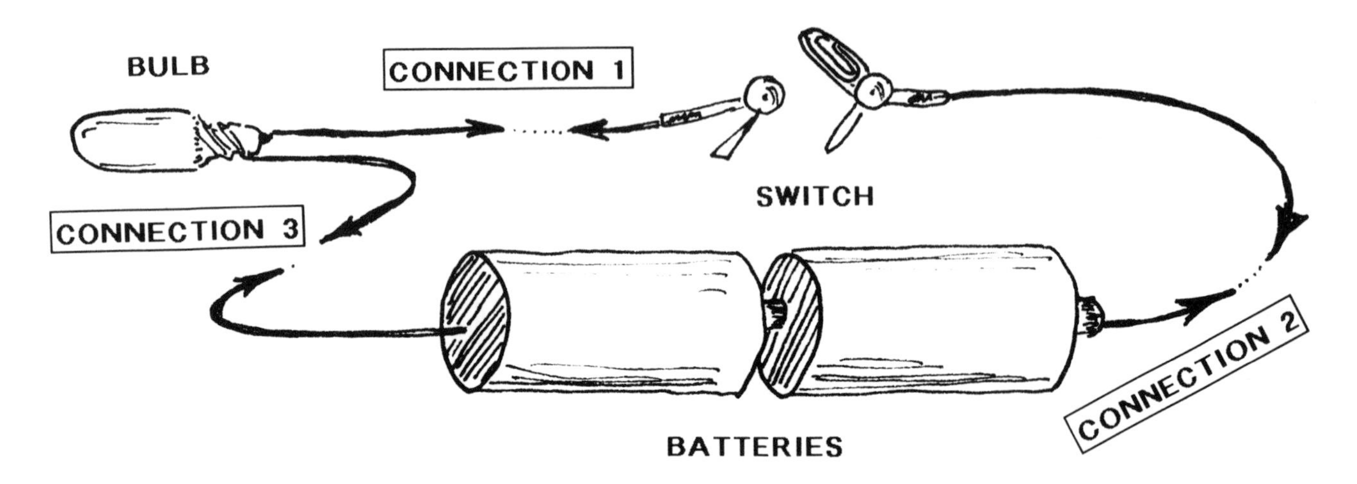

Inasmuch as it is difficult to solder inside the two-liter container, wires are crossed and twisted in a tight pigtail manner. To insure their staying joined, attach small, terminal wire connectors. These are twisted on the joined, wire ends until tight. Turn on the paper-clip switch. Your flashlight bulb should be illuminated. If not, check the bulb. Check the batteries. And check all your wire connections.

THE TWO-LITER ZOETROPE

An image will remain on the retina of the eye for only one-sixteenth second after the object has been removed from sight. This phenomenon is called persistence of vision or retinal retention. When the eye is subjected to a succession of images at rapid speed, the brain combines them into one moving picture.

To construct your two-liter, plastic container Zoetrope, you will need a Manila folder, a two-liter container bottom section, two wooden beads, a six inch length of one-eighth inch, wooden dowel, glue, and a pair of scissors.

HOW TO CONSTRUCT A ZOETROPE

Remove the bottom section from a two-liter container. This can be accomplished by cutting a two-liter plastic container in half. Apply the heat from a hair dryer to the inside of the container directing the heat to the base. The bottom section can now be easily separated from the container. Using one of the "super" glues, glue an approximately, three-eighths wooden bead in the middle of the inverted bottom. Wooden beads can be purchased at most craft stores. Let the glue dry overnight.

From Manila folders, cut several, thirteen inch long, by one inch wide strips. To avoid the Manila folder crease and to obtain the thirteen-inch length, measure and cut these strips from the diagonal of one half of a folder. When measuring, include portions of the folder to act as a tab for future overlapping and joining.

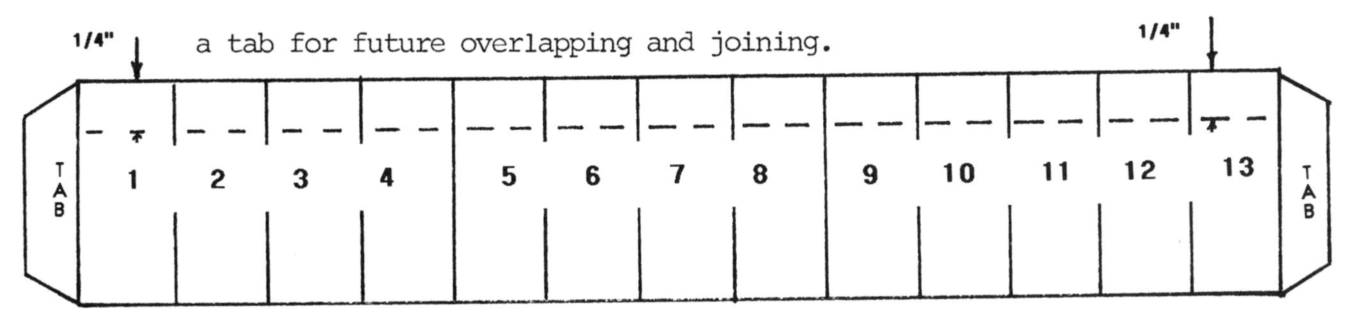

Draw in thirteen, one-inch wide frames. Draw a line one-quarter of an inch from the top and parallel to the longest dimension of the strip.

Wrap this strip around the interior base of the two-liter bottom section. Anchor the strip evenly to the bottom section's edge using several paper clips to hold it in place. Avoid any shifting.

Using a one-hole, deep-throated, paper punch and working from the inside out, punch holes at all the "X" intersections. Once all the holes have been punched, maintain alignment of these holes. For this purpose a few golf tees, placed here and there through the holes in the bottom section and the Manila strip, work well.

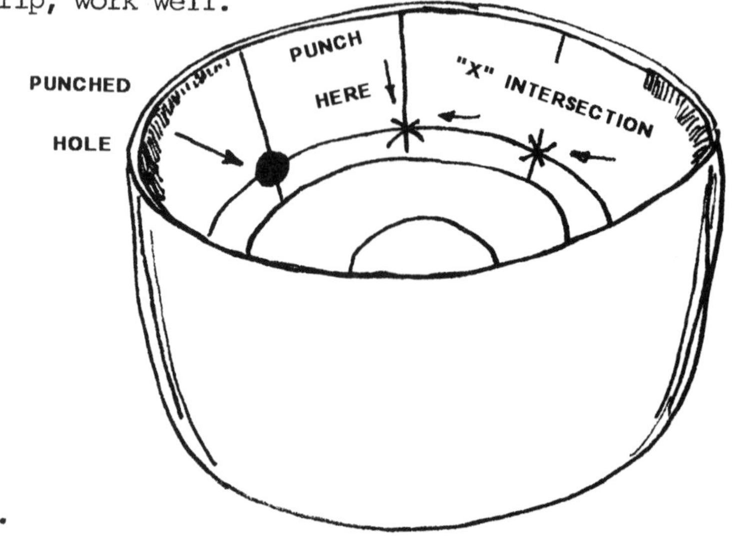

With a pair of scissors, make parallel cuts up from the bottom reaching up to the punched holes. You should be cutting through the plastic and the Manila folder strip underneath simultaneously.

Carefully remove the strip. Decide on a topic for your moving picture. This could be a person sneezing, a bird flying, a tree falling to the ground, a pendulum in motion, or a flower opening and closing, or anything else that moves. To best sequence your frames depicting motion, for example a falling tree, start out with frame one and show the tree standing and then complete frame thirteen showing the tree resting on the ground. At mid-point in your strip, frame seven, show the tree inclined at an angle of forty-five degrees or half way down to the

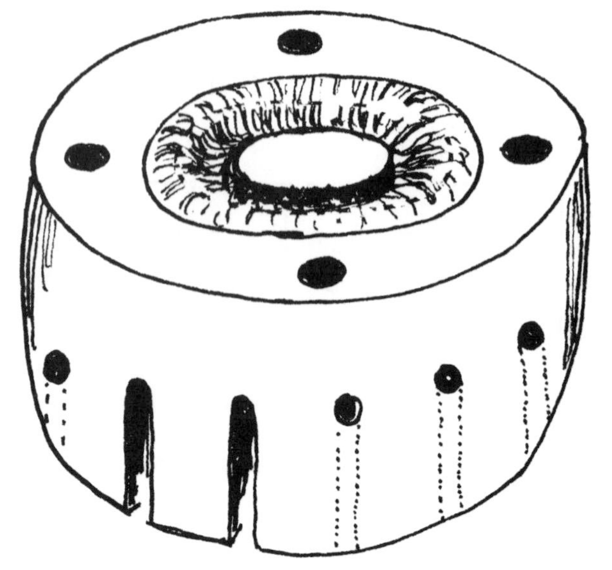

ground. These three, completed frames guide you in filling in the remaining frames in an orderly sequence. Map out your selection on a piece of scrap paper. When you are certain as to what you wish to portray, draw in each frame on the Manila strip.

When you have finished framing your selection, for example, the pendulum,

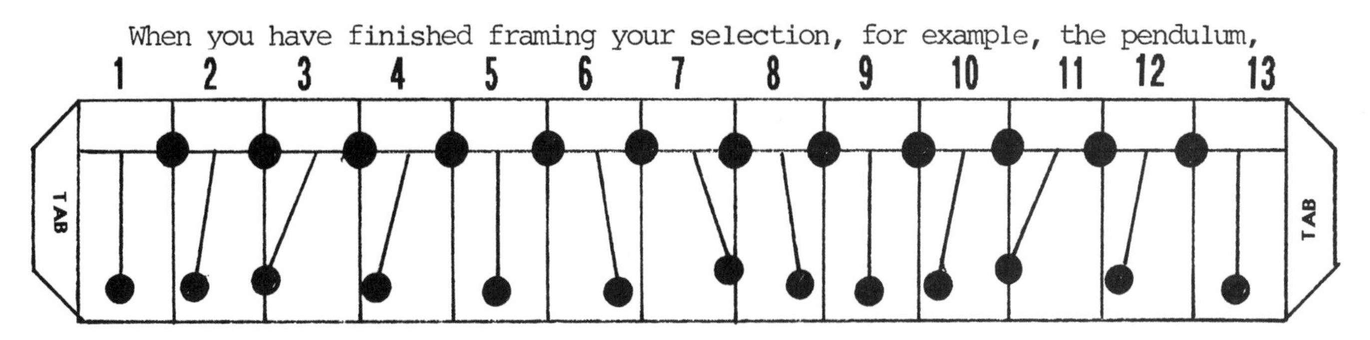

insert the strip back into the interior of the two-liter bottom section. Carefully place each frame so that the holes in the Manila folder strip align themselves with the holes cut in the plastic. Once this is done, use snippets of clear tape to lock it in place.

To the six inch length of one-eighth inch, wooden dowel, glue a three-eighths, wooden bead on one end of the dowel. Let this dry thoroughly. Smooth off the opposite end of the dowel. If dowels are not available, a six inch length of clothes hanger wire will work nearly as well. Insert the dowel (or wire) into the wooden bead previously cemented to the interior top of the two-liter bottom about the vertically held dowel. Observe the motion of your selected topic drawn on the Manila strip. Try a variety of speeds.

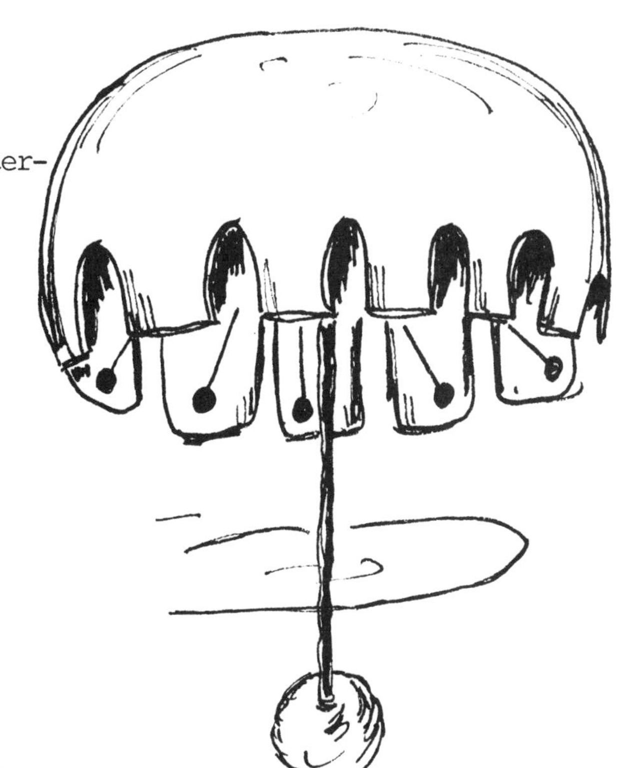

THE TWO-LITER STRING TELEPHONE

You will need two, two-liter plastic containers devoid of their bottom sections and a twenty-five foot length of heavy thread that will not stretch.

Cut two, two-liter containers in half. Discard the top portions. Remove the bottom section from each container. From the inside of the half sections of both containers, carefully pierce the center of the bottom portion. Do this with the smallest hole possible to accomodate passage of the string through it. Knot the string about a section of a wooden match so that the string cannot slip out when drawn taut. Thread the other end of the string through the other container. Knot this about a section of wooden match.

To operate your two-liter, string telephone, two individuals must tightly stretch the string. Do not let the string touch anything.

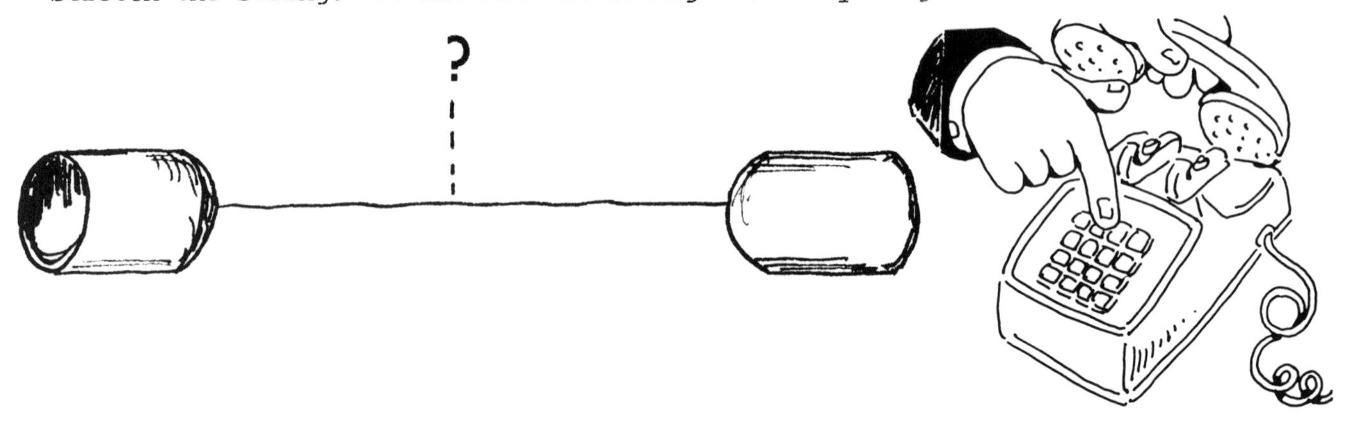

One individual should speak into the cylindrical, sending cup while the other individual holds the cylindrical, receiving cup directly over one's ear and covers the other ear to drown out outside noises.

How well was the message received?

***** Can a string telephone "party" line work?

With one taut line between two individuals, introduce another string, half way between these two individuals and stretched out at a ninety degree angle.

***** Can a fourth individual become another "party" line member? How would you design this telephone hookup?

***** How great a distance can a string telephone carry a message?

When one individual speaks into the two-liter, cylindrical cup, sound waves cause a vibration of the sending cup which is then transmitted to the receiving cup.

THE TWO-LITER BIRD HOUSE AND BIRD FEEDER

Two-liter, plastic containers are a natural for conversion into bird houses and bird feeders. The containers are waterproof, holes can easily be made in them, and dowel-type, bird perches can readily be inserted into the containers.

Check a reference book about birds. Make a selection of one bird to which you wish to cater. Construct a feeder that conforms to the variety of food that a particular bird prefers, the food opening must match the bird's beak structure and size. An incorrect opening may make it impossible for the bird to reach the food. Or, if it is too large, you may be inviting numerous unwanted guest and the bird you wish to feed may get little or no food.

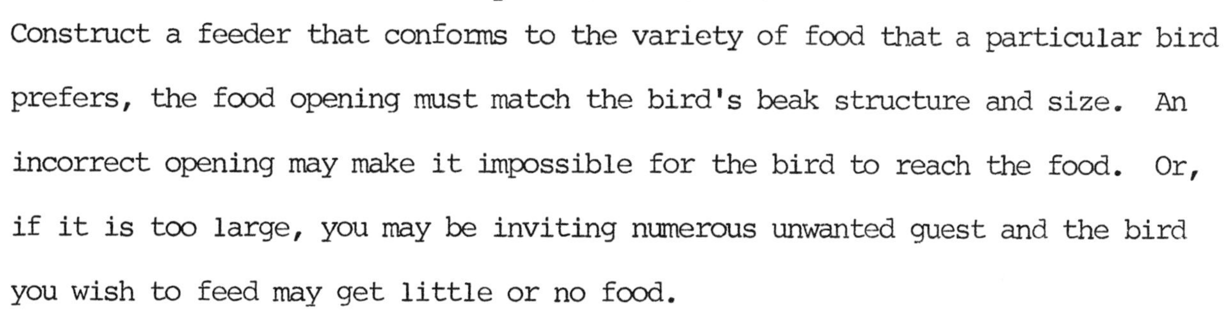

Bird houses should not be made of clear materials. Birds like privacy. They usually prefer a home that is natural in color and appears weather worn rather than spanking new and shiny. You may wish to paint containers that you use for bird houses. Also, the opening to the bird's house should match the size of the bird's body and it should be hung in a spot to his liking. The location of your bird feeder or bird house must be a safe place, quiet, and readily visible from the air. Specific birds have specific preferences. A reference check will inform you of an individual bird's likes and dislikes.

A hole punched into the cap of the container provides a means to suspend your feeder or bird house. Thread the string or fishing line through the hole in the cap and knot the string around a small piece of twig to keep the string from being pulled out under pressure.

THE TWO-LITER ICE ROLLERS

This activity utilizes three, two-liter, plastic containers. Fill each container with one liter of water. Cap them. Position these containers in a freezer so that the water can freeze in three different positions, for example, vertical, horizontal, and inclined.

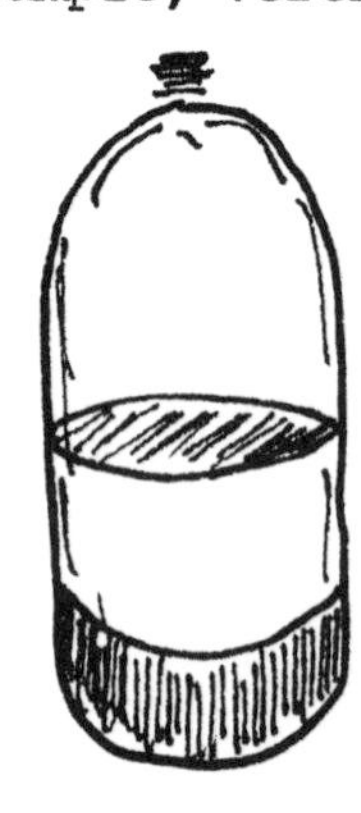

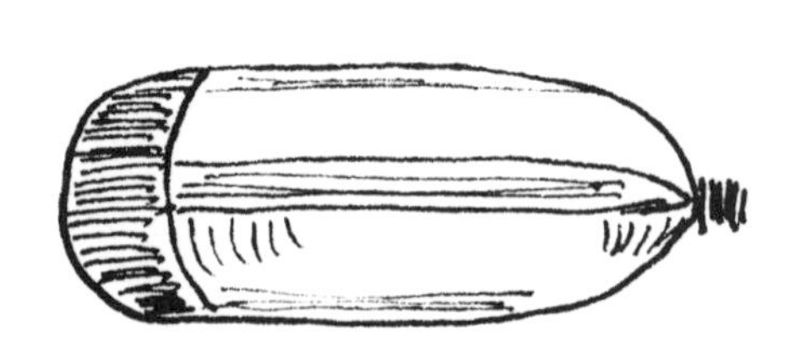

***** With equal amounts of water frozen in three different orientations, infer what the performances would be as each is released down an inclined plane. Will they all travel at the same speed down the inclined plane? Will they all travel in a straight path down the inclined path? Will they all travel the same distance out from the end of the inclined plane?

***** Given another two-liter container and one liter of water, what new design can you fabricate for the internal frozen water? What results do you infer your design will provide?

***** Fill each of three, two-liter containers with one liter of water. With all containers positioned vertically, freeze the water. When this is accomplished, keep one container capped and uncap the remaining two. In one of the uncapped containers place a small, Manila folder, block-letter "T" cutout. Infer and rank order the order of melting, fastest to the slowest. Did your inference match your observations?

CAPPED

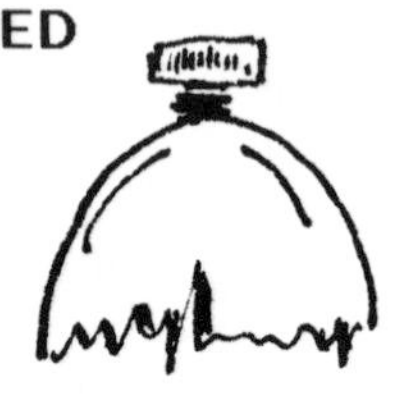

UNCAPPED

BLOCK "T"

INDEX

TWO LEADERS of SCIENCE INSTRUCTION